JAMES GREEN

CANAL BUILDER AND COUNTY SURVEYOR

JAMES GREEN

CANAL BUILDER AND COUNTY SURVEYOR

(1781–1849)

BRIAN GEORGE

DEVON BOOKS

First published in Great Britain in 1997

Copyright © 1997 Brian George

All rights reserved. No part of this publication may be reproduced, stored in a retrieval system, or transmitted in any form or by any means without the prior permission of the copyright holder.

British Library Cataloguing-in-Publication Date
A CIP record for this title is available from the British Library

ISBN 0 86114 914 9

DEVON BOOKS
Official publisher to Devon County Council
Halsgrove House
Lower Moor Way
Tiverton EX16 6SS
Tel: 01884 243242
Fax: 01884 243325

Front cover illustration: Cowley Bridge, Exeter, photographed in September 1997 by David Garner for Devon County Council.
Back cover illustration: Rolle Canal Acqueduct – a 19th century print.
(By kind permission of the Devon and Exeter Institution)

Printed and bound in Great Britain by Devonshire Press Ltd, Torquay

CONTENTS

FOREWORD by Edward Chorlton		5
INTRODUCTION		7

CHAPTER 1 1781–1808
THE EARLY YEARS

1.1	Parentage and place of birth	9
1.2	Assistant to John Rennie	9
1.21	Chelmer and Blackwater Navigation	10
1.22	Totnes survey	10
1.23	Cattewater survey	10
1.3	Land Reclamation for Lord Boringdon	11
1.31	Chelson Meadow	11
1.32	West Charleton	12
1.4	Fenny Bridges	12
1.5	County Bridge Surveyor	14
1.6	The County Administration in 1808	14

CHAPTER 2 1809–1820
BRIDGES, LAND RECLAMATION AND ARCHITECTURE

2.1	The Next Twelve Years	17
2.2	The County Bridges and the County Surveyor	18
2.21	Devon Quarter Sessions Administration	18
2.22	Bridges renewed and repaired	24
2.3	Bridge Work for Other Authorities	37
2.31	Bideford Long Bridge	37
2.32	The River Bridges at Exeter	37
2.4	Crediton and Torrington Canals	39
2.5	Land Reclamation	39
2.51	Braunton Marsh	40
2.52	Budleigh Salterton	41
2.53	John Rennie and the Royal Navy	41
2.6	Elmfield House	42
2.7	Other Architectural Works	44
2.71	Buckland House, Buckland Filleigh	44
2.72	St David's Church	45
2.73	Fursdon House, Cadbury	45
2.74	Killerton House	46
2.8	Family Matters	47
2.9	Turnpike Road, Exeter to Plymouth	47
2.10	The Bude Canal	50
2.11	Towards the next decade	51

CHAPTER 3 1821–1830
BRIDGES, CANALS AND TURNPIKES

3.1	Introduction	53
3.2	The County Bridges and County Surveyor	54
3.21	Devon Quarter Sessions Administration	54
3.22	Bridges renewed and repaired	58
3.23	The County Buildings	69
3.3	Canals	72
3.31	Exeter Canal	73
3.32	Bude Canal	78
3.33	Torrington Canal	80
3.34	Liskeard and Looe Canal	81
3.35	Newton Abbot Canal proposal	83
3.36	Western Ship Canal proposal	83
3.4	Turnpike Roads	84
3.41	Pocombe Bridge to Tedburn St Mary	85
3.42	Countess Wear Committee of the Exeter Turnpike Trust	88
3.43	Proposed Road near Lyme Regis	89
3.44	Flying Bridge over the River Plym to Lady Down near Ugborough	89
3.45	Roads around Stratton	92
3.46	Ilminster Turnpike Trust	93
3.47	Chard Turnpike Trust	94
3.48	Bridges on the new Exeter to Barnstaple Road	95
3.49	Proposed Highway, Topsham to Exmouth	97
3.5	Earthworks	97
3.51	Blachford Park, Cornwood	97

3.6	Harbours	98
3.61	Survey of St Ives and Ilfracombe Harbours	98
3.62	Combe Martin Harbour proposal	98
3.63	Bridport Harbour survey	99
3.64	Cardiff Docks proposal	99
3.7	Residences	100
3.8	The Institution of Civil Engineers	100
3.9	Towards the next decade	101

CHAPTER 4 1831–1841
BRIDGES, CANALS AND NEWPORT DOCK

4.1	Introduction	104
4.2	The County Bridges and the County Surveyor	105
4.21	Devon Quarter Sessions Administration	106
4.22	Bridges renewed and repaired	112
4.23	The County Buildings	123
4.24	Barnstaple Bridge Trust	126
4.3	Water supply, sewerage and railway proposals for Torquay	127
4.4	Canals and Docks	127
4.41	Grand Western Canal	127
4.42	Chard Canal	132
4.43	Burry Port and the Kidwelly and Llanelly Canal	134
4.44	Survey for an alternative London to Birmingham Canal	135
4.45	Stourbridge Canal Extension	135
4.5	Land Reclamation, Westmoor, Somerset	135
4.6	Exeter Turnpike Trust	136
4.7	Residences and bankruptcy	137
4.8	The Institution of Civil Engineers – Green's First Paper	138
4.9	Newport Dock – dismissal by Devon Quarter Sessions	142

CHAPTER 5 1842–1849
WESTMINSTER, LONDON

5.1	Introduction	144
5.2	Completion of Newport Dock	145
5.3.	South Devon Railway Bill 1844	148
5.4	The Institution of Civil Engineers – Paper on the Exeter Canal	148
5.5	Joseph Dand Green	149
5.6	The Institution of Civil Engineers – Drainage and Sewerage of Bristol	150
5.7	His death	151
5.8	Review of Green's achievements	151

APPENDICES

A The Report of James Green on the Devon County Bridges, 10 January 1809	153
B List of Devon County Bridges, 1831	161
C Specification of Works in Repairing Bridges, 1831	165
D Bridge design and construction ascribed to James Green, 1808–1841	167
E James Green's canal and dock proposals and works	170
F Membership of the Institution of Civil Engineers 1838	171

BIBLIOGRAPHY	177
REFERENCES	178
ILLUSTRATION ACKNOWLEDGEMENTS	182
INDEX OF NAMES	183
INDEX OF SITES	185

Note: The captions to photographs of bridges give as appropriate, the date of Green's construction and a National Grid Reference for location.

FOREWORD

Throughout history transport has had a high profile whether it be helping Hannibal cross the Alps or in the current realisation that we must make a fundamental change of direction if we are to adjust the balance between our over-reliance on the car and the need to develop more integrated and sustainable transport systems.

One constant in this period of rapid change will be the fact that people still need to travel between cities, towns and villages which have functioned as communities for centuries, in many cases reaching back to the Middle Ages. The highways system which will take us into the new millennium will be very much based on the heritage of road, railways, bridleways and footpaths which has been built up over many generations. As we travel through our towns and countryside, we are following in the footsteps of our ancestors.

Despite this country's long engineering heritage, many of the personalities who pioneered its design and construction have escaped the public eye. Compared with artists, politicians and other professions, far too few eighteenth and nineteenth century civil engineers are remembered as household names today.

I remember many years ago being inspired by Samuel Smiles' classic book *Lives of Engineers* and by L. T. C. Rolt's very accessible biographies of Isambard Kingdom Brunel, George and Robert Stephenson and Thomas Telford. The latter, another inspirational factor in my own career, was the County Surveyor of Shropshire, the county of my birth. Some of his early work was as engineer to the Ellesmere Canal Company and he became the first President of the Institution of Civil Engineers in 1820.

In 1800 there was correspondence between the Clerks of the Peace of Devon and of Shropshire concerning the appointment of a County Surveyor for Devon, to be responsible for its bridges. The correspondence explained Mr Thomas Telford's role as County Surveyor in Shropshire and helped Devon to come to the view that it should appoint a County Surveyor with responsibility for inspecting and reporting on 237 of the county's bridges, bringing them up to a good standard.

James Green was appointed to this new position in 1808. Although he has many mentions in *Devon Roads* (published by Devon Books in 1988), his achievements deserve wider recognition and Brian George's new biography is to be warmly welcomed.

James Green – Canal Builder and County Surveyor

Green started his career at the beginning of the nineteenth century assisting the great John Rennie on schemes such as the Plymouth Breakwater. He went on to establish a reputation in his own right designing a remarkable range of civil engineering projects – docks, land reclamation schemes, houses and churches as well as transport projects such as roads, canals and bridges. It was perhaps his misfortune that his professional career developed in the years before the growth of the railways but many of his structures still survive today and a good number are listed as buildings of special architectural or historic interest.

I welcome this book for three particular reasons. James Green was appointed Devon's first County Surveyor and, as the current County Environment Director for Devon, I am well aware of the engineering legacy he has left in this County and which still serves the needs of its residents and visitors today. Many of the 237 bridges for which Green took responsibility in 1808 still stand 190 years later. Secondly, the story of Green's life set out in this volume is based to a large degree on historical records held in the County Record Office in Exeter – for example the very fine engineering drawings preserved in the Victorian Bridge Books demonstrate the quality of many of Green's bridge designs. One of the aims of the County Council's Devon Books imprint is to bring such records before a wider public and this book serves as an indication of the wealth of the County Council's archives.

Lastly, this biography gives me great pleasure as it has been written by a former member of Devon County Council's engineering staff. I continue to be impressed by the depth and knowledge held by County Council officers, past and present, and Brian George's research is yet another example of this tradition. I feel confident that his book will help readers better appreciate the skill that lies behind the engineering heritage of Devon and of the country at large.

Edward Chorlton
County Environment Director
Devon County Council, October 1997

INTRODUCTION

No one who has taken an interest in the construction of canals in the south west of England can be unaware of the name of James Green. Charles Hadfield in his book on these canals, published by David and Charles in 1967, includes Green in his list of names to whom the book was dedicated. Two further books on Green's work on canals, published about five years later, covered the subject in much more depth. These were *The Bude Canal* by Helen Harris and Monica Ellis and *The Grand Western Canal* by Helen Harris. The latter book has recently been revised, updated and republished.

But it was Michael Dickinson of the Devon County Record Office, in his unpublished *Appreciation of James Green and the Devon Quarter Sessions*, who whetted my appetite to know more of this man, because Dickinson exposed the vicissitudes of Green's life from his examination of the Quarter Sessions records. Green's work for the Quarter Sessions as County Bridge Surveyor, later to be called County Surveyor when he took responsibility for the county buildings, was the mainstream of his civil engineering career even if it is for his canal work that he is more widely known.

This book therefore describes his county work in detail, quarter by quarter, together with all those other activities that he pursued in his other roles sometimes as consultant or as contractor. As his career progressed those roles sometimes seemed to compete for his best attentions, and for over forty years his thoughts encompassed canals, land reclamation, architecture, bridges, docks and highways – a wide range of expertise indeed.

For the information in Appendices B and C concerning the 1831 bridge list and the maintenance contract I am greatly indebted to Dr G.L. Cantrell who allowed me to view the original documents, material which has since passed to the museum of the Worshipful Company of Pavoirs; also to Frank Oates for passing to me details of Green's work for Lord Boringdon.

My former colleague, David Thomas, has devoted years to the study of the history of Devon bridges and I am particularly indebted to him for the information he has so freely fed to me from time to time. The existence of a computerised list of county bridges has been most helpful and I am grateful for the support given to me by Alfred Cornish, Malcolm Bramley, Colin Hatherley and Peter Warren of the county bridges office.

John Bentley and David Greenfield of the Somerset Industrial Archaeological Society have contributed much information on Green's work in that county. Owen Gibbs of Cardiff has provided information on the proposed 1829 Cardiff Docks and the 1842 Newport Dock.

I have always been aware of the enthusiastic help afforded by successive librarians of the Institution of Civil Engineers, Miss D. J. Bayley and Mr M. M. Chrimes and the secretaries of its Panel for Historical Engineering Works, Mr W. A. Morris and Mrs. M. K. Murphy, the archivist; also the encouragement of the late Professor Walter Minchinton, Chairman of the Exeter Industrial Archaeological Group.

As a former member of staff of the Devon County Council, I am very pleased to be able to note the excellent facilities and the enthusiastic help of the staff of the West Country Studies Library and the County Record Office in Exeter; indeed librarians everywhere show their committment to their professional work.

For those who wish to seek out the bridges mentioned in this book the full national grid reference is given in Appendix D. A journey across Devon can be planned to pass over chosen bridges but please take care to avoid parking your vehicle in a dangerous position. Remember that most bridges are narrow by today's standards and great care must be exercised when on foot between the parapets. If examining the bridge from the river bank beneath, however safe it appears to be at first glance, ensure that you have someone with you in case of an accidental slip. This is the precaution that professional bridge inspectors take.

CHAPTER 1
1781-1808
THE EARLY YEARS

1.1 PARENTAGE AND PLACE OF BIRTH

Although the name of James Green is little known in Devon, he did much to improve transportation in the county before the advent of the railway. He applied the knowledge gained by people such as James Brindley, John Smeaton, Thomas Telford and John Rennie earlier in the eighteenth century, and while he was resident in Exeter for 33 years Green also undertook works in Cornwall, Somerset and South Wales.

James Green was born in Birmingham in 1781. His father was a civil engineer and also a contractor in Warwickshire and the adjoining counties, whose name was probably also James Green, and it was from him that James received his early experience until in 1801 he was employed by John Rennie.[1]

The name of James Green, at dates which might have been suitable for the work of the father, arise twice in canal history as reported by Charles Hadfield and S. R. Broadridge.

Firstly, William Jessop was asked to survey a navigable cut from the Cromford canal to the Trent, but he fell ill and suggested that James Green, a surveyor who worked for Lord Middleton at Wollaton, should do the work instead. The committee agreed to this in June 1791, provided Green worked under Jessop's direction which he did as a resident engineer until 1796. From 1792 Green also assisted Jessop as engineer for part of the Grantham canal from the Trent to the Leicestershire boundary until completion in 1797.[2,3]

Secondly the construction of the Stourton to Stourbridge canal and the Stourton to Dudley canal were begun by committees who shared their clerk and engineer. In 1801 the Dudley canal committee appointed James Green, engineer of the Stourbridge canal, as engineer to the Dudley canal, each canal paying him £35 per annum.[4]

1.2 ASSISTANT TO JOHN RENNIE

Rennie was the great engineer who had conceived a lighthouse on Bell Rock off the Firth of Tay and who was to build the Plymouth breakwater for the Admiralty. His Kennet and Avon canal from Bath to Reading joined London with Bristol but his Grand Western Canal from Tiverton to Lowdwells on the Devon/Somerset border was only part of a project never realised of joining the Bristol and English Channels. His iron bridge over the Wye at

9

Chepstow survives but his great masonry bridges over the Thames have been replaced as traffic has increased. Waterloo Bridge, built in 1817 was the first, and his design for London Bridge was built by his son after his death.

James Green worked as assistant to Rennie from 1801 to 1807, leaving his service after a year in Devon. Rennie had employed Green 'on extensive surveys, canal works, the drainage of bogs and fens and the design of engineering works generally, both in England and Ireland. At that time the repair and replacement of Dymchurch Wall came particularly under Green's care and the reconstruction of the sea lock of the Chelmer and Blackwater Navigation was entirely trusted to him by the Earl St Vincent.'[1] This lock was 107 feet by 26 feet giving access to vessels drawing 8 feet at neap tides and 12 feet at springs. It has been lengthened since the Second World War.

1.21 CHELMER AND BLACKWATER NAVIGATION

Before the younger James Green joined John Rennie, Rennie had reported to Lord St Vincent regarding some criticisms by the Chelmer and Blackwater Navigation Committee on 5 June 1799.[5] A further report of 4 December 1805 discussed compensation for certain mills having a reduced supply of water and mentioned that 'the lock as measured by Mr Green is 69 feet long, 17 feet wide and 9 feet 2 inches rise containing 10,772 cubic feet'. In a final paragraph Rennie showed that Green was trusted as his representative for the works:

The above are the material alterations which appear to me requisite to the improvements of the Chelmer Navigation – there are no doubt several other works which will require to be done to render it complete to Chelmsford but they are of such a Nature as may be accomplished in the common routine of repairs, and therefore need scarcely be enumerated in this Report. I would however advise that Mr Green be directed to go over the whole carefully and state to your Lordship for the information of the Committee, the particulars of each of the Expenses that would attend them.[6]

1.22 TOTNES SURVEY

It was probably from there that Green came to Devon for, while in November 1805 Rennie was giving a surveyor, Charles Tozer, instructions on a survey of the river Dart at Totnes on behalf of the Duke of Somerset, in July 1806 Green was doing likewise.[7] Rennie reported to the Duke of Somerset on ways of improving the navigation below Totnes Bridge to the Quay at Fleet Mill Reach or of providing a canal and quay on the Bridgetown side of the river.[8]

1.23 CATTEWATER SURVEY

Rennie also employed Green on a survey of the rock at Cattewater intended for use for the construction of Plymouth Breakwater, which scheme was commenced in 1812.[9] On 11 August 1807 Rennie wrote to the surveyor general of H.M. Land Revenue 'considering a plan and report of Mr Green for laying mooring chains in the harbour of Cattewater.'[10]

1.3 LAND RECLAMATION FOR LORD BORINGDON

Meanwhile in a report to Lord Boringdon of Saltram on 30 December 1805 regarding an embankment from Pomphlett Point to Saltram Quay, Rennie had proposed two alternative schemes and drawn a section of the proposed embankment 76 feet wide at the base and 13 feet high above the estuarial mud which he estimated would cost £4,320.[11]

1.31 CHELSON MEADOW

This obviously had a favourable reception for in the spring of 1806 Lord Boringdon contracted with Green for the construction of an embankment to enclose 175 acres of land formerly called Chelson Bay in the estuary of the river Plym.[12] This embankment was 970 yards long, 91 feet wide at the base and of average height 16 feet above the estuarial mud. The outer slope of 1 in 4 gradient was covered with limestone set on a watertight puddle of clay 3 feet thick. Green did his own research into the height of the tides and in the latter stages of construction raised the bank 2 feet above the planned level.

The additional height and consequent additional width of the embankment caused a collapse during construction due to the greater weight resting on the estuarial mud. Green therefore changed the batter on the estuary face for the last 3 feet of height from 1 in 4 to 1 in 2. There were other problems. On 15 October 1806 James Green complained to the Mayor of Plymouth that two men employed on Lord Boringdon's embankment had been impressed by the Royal Navy.[13]

Embankment for Chelson Meadow 1807   *(SX505544 – 507552)*

1.32 WEST CHARLETON

For this work and for reclaiming 40 acres from a small inlet to West Charleton in the Kingsbridge estuary, Lord Boringdon received a gold medal from the (Royal) Society of Arts for his enterprise. The cost was about £9000.[12] Chelson Meadow has since been used for agriculture, as a racecourse, as a First World War airship substation for the RNAS and, until 1929, as a flying field.[14] In recent years it has been filled by controlled tipping of Plymouth refuse.

1.4 FENNY BRIDGES

With Lord Boringdon's assignments completed by November 1807, and although he had been working at Cattewater in the summer for Rennie, there was presumably little work to come from Rennie in Devon as the breakwater scheme was five years away. Green was 26 years of age and must have been looking for an opportunity to better his position – or perhaps he wished to stay in Devon! His success meant that he was poised to undertake further civil engineering projects in the county and his opportunity came as the direct result of the collapse of the newly rebuilt Fenny Bridges in November, 1807.[15] These bridges carried the London to Plymouth road across the river Otter, west of Honiton.

In April 1808 he contracted for the design and construction of a replacement bridge on this important road. He built a three-span arch bridge of 42, 48 and 42 feet spans in brickwork with a roadway 20 feet between parapets. This has carried main road and trunk road traffic ever since, though displacement of the parapet copings is a constant reminder of the change in the type of traffic over 190 years. A subsidiary 40 feet single-span arch across a small stream on the western edge of the same flood plain was lost in the East Devon floods of July 1968 and has been replaced by a more suitable structure.[16]

When Green signed his contract for work on Fenny Bridges he was asked to do some repair work on other bridges in East Devon at Ottery St Mary and Newton Poppleford, though this was to be under supervision. His reputation spread and at a meeting of the Plymouth Eastern Turnpike Trustees on 3 October 1808 it was reported that bridges over the river Yealm (at Lea Mill) were too narrow and that plans placed before the meeting by Mr Green 'appeared to be well calculated to remove the evils complained of'. Mr Harry Woolcombe was sent off to the Quarter Sessions next day to present the

Fenny Bridges 1808   *(SY114985)*

THE EARLY YEARS

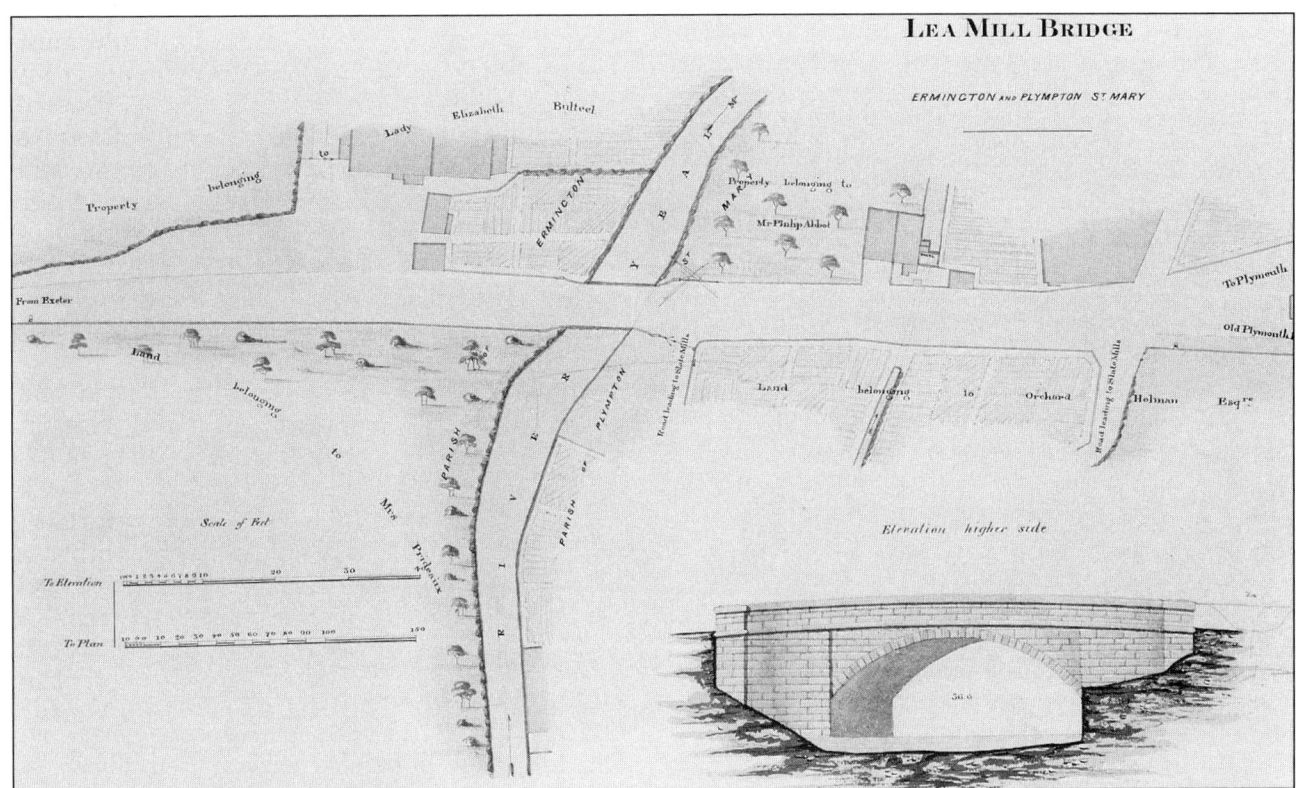

Lee Mill Bridge, plan and elevation 1808 (SX599557)

plans to the magistrates and a contribution of £500 towards the cost was voted by the county. The bridge over the river Yealm is particularly interesting because it was built to a skew of 21°. It is an early example of a skew bridge and an indication of the confidence of the engineer in constructing a sound structure in this manner.

Meanwhile, during the Midsummer Sessions of 1808, a committee of magistrates had examined the current practice of looking after the county bridges. They had available correspondence between the Clerks of the Peace of Devon and Shropshire in July 1800, when the latter Clerk had provided information on the conditions of the appointment of Thomas Telford as surveyor of bridges for that county.[17] The Devon committee reported that the present system was inadequate, that the number of bridges to be maintained at the expense of the county was about 250 and that the whole ought to

13

be inspected by the Surveyor at least once a year though it was not necessary that the civil engineer should be confined to county business. Furthermore, a person properly qualified should be procured at £300 per annum as the present six surveyors cost £145 annually, and also, this person should furnish all plans and estimates gratuitously and supervise the erection of the new bridges.

The order to issue advertisements to the press and to answer inquiries giving the number of bridges and the nature of the duties required in Devon was given on 22 July and at the Michaelmas Sessions on 4 October 1808 James Green was appointed.

1.5 COUNTY BRIDGE SURVEYOR

James Green was Devon's first County Bridge Surveyor, a title which was quickly to become County Surveyor and which was held by his successors until 1888. Then the turnpikes ceased to exist and the County Councils were formed to administer both the county roads and bridges and the term County Surveyor was given to the county's chief officer who was responsible for these combined functions.

After 1888 Devon went through a period when it was unsure whether to have one, two, three or four County Surveyors responsible to the County Council for the eastern, western, southern and northern parts of the county or a combination of any of these. In 1947 the title of County Surveyor was assumed by Mr R.B. Carnegie when he took responsibility for the whole of Devon's highway and bridge construction and maintenance. When Mr Carnegie's successor, Mr Henry Criswell retired in 1974, the title was changed to county engineer to reflect additional duties exercised by the chief officer and later the title became county engineer and planning officer. In 1994 the title was changed to county environment director.

1.6 THE COUNTY ADMINISTRATION IN 1808

Devon administration in 1808 was largely conducted by the Quarter Sessions, the parishes, the municipal boroughs and the relatively new turnpike trusts. Quarter Sessions, primarily responsible for law enforcement, employed a Clerk of the Peace, County Solicitor and County Counsel, keepers of the various prisons, a County Treasurer and either a County Surveyor or various surveyors to manage the county bridges.

Parishes employed constables, overseers of the poor and surveyors of highways. These people were principal inhabitants of the parish who served for a year and then hopefully passed their duties to another parishioner. They were not professionals. The municipal borough organisations were separate from the county.

The turnpike trustees looked after about ten per cent of the road mileage in the county and these were the main roads from Cornwall to Somerset or between the Devon towns.

Bridges were administered by all four authorities. Small bridges on less important roads were looked after by the parishes, bridges in a borough were the responsibility of the borough. The turnpike trusts built and maintained the bridges under their roads, but the Exeter Turnpike Trust was responsible under its act for the city bridges across the river Exe and the Countess Wear Bridge, while Tiverton Turnpike Trust was responsible under its act for Bickleigh Bridge.

The only bridges for which the county was responsible were those for which they had assumed responsibility under the law. We shall see that in 1808 Devon had responsibility for just over 230 bridges and these were called county bridges. Quarter Sessions devolved the responsibility to the magistrates who lived in the nearby area, grouping certain bridges and magistrates into divisions, presumably as the law operated. The surveyor was responsible to Quarter Sessions but in co-operation with the magistrates of the divisions who he would keep informed and meet when carrying out his formal inspection of each bridge.

At the turn of the century the system for maintaining the county bridges was extremely cumbersome. The surveyor, finding work was necessary to maintain a bridge had to present the bridge at Quarter Sessions and the magistrates assembled would decide which bridges presented would receive attention, always bearing in mind the cost of the work within the county rate. Some presentments would be successful, some would not, and for the successful scheme an order would be made that the bridge would be repaired or rebuilt at a cost not exceeding £X. William White, whose Fenny Bridge was reported fallen down at Quarter Sessions of January 1808 had presented 18 bridges at the January 1803 Sessions; at Easter seven of these were presented again; at Midsummer six of the seven once more and at Michaelmas four of the seven once again. At the January 1805 Sessions White presented 28 bridges, at Easter 18 of these again, at Midsummer 14 of the 18 once more and at Michaelmas 12 of the 14 yet again.[18]

Another use of the system of presentment was to enlarge the stock of county bridges, and as might be expected, the case for a bridge to become a county bridge was always strongly resisted by the court. The evidence for the status of a bridge would be given by the Clerk of the Peace to the court at Quarter Sessions, which would then direct the County Solicitor to enquire into the facts, to take the advice of the local magistrates, the County Surveyor and perhaps the County Counsel. The solicitor would report back three months later at the next Quarter Sessions and the bridge would or would not be accepted by the court. If it were not accepted and the presenter felt sufficiently strongly and had sufficient financial reserves there could be an appeal to the King's Bench as a higher authority.

The turnpike trusts made more and more use of this procedure as the years passed to shed themselves and/or the parishes of the responsibility of rebuilding more capacious bridges to carry the increasing traffic on the more important roads. However, the construction of a new turnpike road on a new route did not necessarily mean that its bridges became county bridges. The turnpike trust did, therefore, often ask the County Surveyor to provide the plan and specification and to superintend the construction of the bridges, as later the County Surveyor could report that they had been built to his satisfaction. This would allow the court to assume responsibility for maintenance under an Act of 1803, while the County Surveyor could charge the turnpike for the provision of the plans and specifications.

This Act of Parliament 43 GIII c59 came into effect on 24 June 1803. There were seven sections and these can be simplified as follows:
1. Surveyors to have powers to get materials for the repair of bridges in the same manner as surveyors of highways under 13 Geo III c78.
2. Quarter Sessions may widen and improve, or alter the situation of county bridges on presentment of their insufficiency.
3. Tools and materials provided by Quarter Sessions are vested in the surveyor.
4. Inhabitants of counties may sue for damages done to bridges in the name of the surveyor. No action or prosecution brought by or against the inhabitants shall be abated or discontinued by the death or removal of such Surveyor.
5. No bridge hereafter to be built in any county at the expense of any individual person(s) or corporate body shall be taken as a county bridge unless erected in a substantial and commodious manner under the direction of the County Surveyor, who is required to superintend and inspect the erection when requested by the person or parties concerned.
6. Special references to the ridings of the county of Yorkshire.
7. Nothing here contained shall extend to any bridge or road which any person or body shall be liable to maintain or repair by reason of tenure or by prescription.

It was into this framework of law and administration that James Green became County Surveyor in 1808. As a former assistant to the renowned John Rennie he came with firm ideas on the need for bridges that would be sufficiently capacious for rivers in flood to pass without scour and sufficiently wide between parapets to enable horse-drawn carriages and wagons to pass each other at speed on the main routes. As time went on it appeared that his ideas were well received by some of his employers but that other magistrates rather resented his forcefulness.

CHAPTER 2
1809-1820
BRIDGES, LAND RECLAMATION AND ARCHITECTURE

2.1 THE NEXT TWELVE YEARS 1809–1820

During the next 12 years James Green set the foundation for his reputation sufficiently for J.M. Rendel, builder of Laira Bridge and later president of the Institution of Civil Engineers, to call Green 'the great engineer in Devon.'

First there were the bridges to be rebuilt and strengthened. Reference to the list (Appendix D) will show how much was achieved in this period. Then there were the consultancy schemes. Quarter Sessions allowed him to seek work from others than the county of Devon and in a series of similar advertisements in the *Exeter Flying Post* this one appeared on 10 November 1808.

> James Green, Surveyor and Civil Engineer begs leave to inform the noblemen and gentlemen of Devonshire and the adjoining counties that in consequence of his recent appointment as general surveyor of the Devon county bridges he has taken up his residence in the city of Exeter, and respectfully solicits their patronage in the several branches of his profession.

Thus followed in 1810 the widening of Bideford Bridge for its trustees and, in conjunction with his visits to rebuild Hele Bridge, Hatherleigh, Green supervised re-building work at Buckland House at Buckland Filleigh. At the same time he was constructing his own home at Elmfield in Exeter. Next year Green commenced the Crediton canal, but this work was almost immediately suspended. He was searching for a route for the Torrington Canal on the north-eastern side of the Torridge but this also came to nothing. However, in September 1809 he had made proposals for reclaiming Braunton Marsh and this work was proceeding satisfactorily. Also in 1811 he made recommendations to the Exeter Turnpike Trust regarding the training of the course of the river Exe on the upstream approach to the bridge it had built in 1778. In 1812 John Rennie wrote to Sir Thomas Acland at Killerton offering his and Green's expertise in the construction of a water supply for the house.

Sometime about 1813 Green commenced the reclamation of land from the estuary of the river Otter at Budleigh Salterton, started some architectural work at Fursdon House near Cadbury and was writing in support of John Rennie in a small matter concerning the tideway at Gosport. In 1813–14 he was allowed to contract with the City of Exeter and County of Devon to rebuild Cowley Bridge and this ancient monument is probably his finest structure. He lost an infant son at Christmas 1815 but another son was born in September 1817 and Joseph grew up to assist his father in his work in the 1830s. He received the commission to rebuild St David's church in Exeter in 1816 and although the church was rebuilt again at the turn of the century its distinctive appearance is often noted in local writings.

In 1819 his consulting work once again turned to civil engineering and he reported to the Exeter, Ashburton–Buckfastleigh and Plymouth turnpike trustees on the realignment of the Exeter to Plymouth road and 14 miles of new road were eventually built. One of his most important and memorable schemes was started, the Bude Canal with its large reservoir on the upper reaches of the Tamar and 35 miles of canal to Holsworthy and Druxton near Launceston.

His bridges and other works extended the length and breadth of Devon and involved him in constant and extensive journeys from his home in Exeter. He would have had to be an excellent horseman and a remarkably fit and resourceful person with considerable organisational ability. By 1820 he had transformed the condition of the Devon bridges by rebuilding many important structures and by establishing the repair of the others to a higher overall standard. At the end of this period, Green's status with the county was enhanced by his appointment to supervise the reconstruction of the sheriff's ward and to carry out work on the castle at Exeter.

2.2 THE COUNTY BRIDGES AND THE COUNTY SURVEYOR

2.21 DEVON QUARTER SESSIONS ADMINISTRATION FROM 1809

In this context we have to remember that under the Statute of Bridges 1530, Quarter Sessions had been given powers to appoint surveyors who were to be responsible to them for the repair, maintenance or rebuilding of those bridges for which the county had responsibility. The Bridges Act 1803 required that any new bridge erected otherwise than by the county should, as a condition to becoming the responsibility of the county, be constructed to the approval of the County Surveyor and this led the courts to order careful supervision by the bridge surveyor. Thus concern about rising costs and the desire for high standards of construction led to the use of surveyors at all stages of building work.

On 6 July 1800, Mr R. Eales, Clerk of the Peace for Devon, wrote to the Clerk of the Peace of Shropshire enquiring how many county bridges were in that county and how they were inspected or surveyed, whether by the magistrates in the different divisions or by a surveyor appointed for that purpose. If it were by the latter, what salary was allowed and did he give up his whole time or was he allowed to carry on and conduct other business. Shropshire replied within four days that Mr Telford was surveyor for bridges and other public works such as the shire Guildhall, two gaols and three houses of correction, that there were 30 bridges, but that the number was rising, and that Mr Telford was paid a percentage on the money expended in building or repairing them or else against a bill for plans, estimates, time, trouble, hire and expenses.

As we have seen, as a result of the collapse of the newly constructed Fenny Bridges (Exeter to Honiton road, over the river Otter), the Quarter Sessions set up an inquiry in January 1808 to ascertain the reasons. Two weeks later the committee recommended rebuilding to a plan submitted by Mr James Green (former assistant to Mr John Rennie) and the Quarter Sessions confirmed that a contract should be entered into immediately and that Green should also do repair work on two other bridges under the supervision of the magistrates. At Easter Sessions the committee was directed to recommend a civil engineer in lieu of the present six surveyors costing £145 annually and to advertise in the Exeter, Sherborne and two London papers. At Midsummer, Quarter Sessions decreed that it was not necessary that the civil engineer who might be approved should be confined to the county business and that his salary should be £300 per annum, and at Michaelmas Sessions James Green was appointed surveyor to commence his duties immediately. By asking for a civil engineer to be their surveyor, the magistrates must have narrowed the field considerably.[1]

At Easter Sessions 1809, Green produced a comprehensive report on the 236 county bridges, giving spans, width between parapets and the condition of the materials of construction.[2] The bridges are listed in Appendix A. He set out a priority list of work to be done and Quarter Sessions responded by recommending that Mr Green should allocate bridges to districts (divisions) and that their approach roads should be repaired after advertisement. Approach roads were the length of 100 yards on either side of the bridge that had to be kept in repair by the county.

Quarter Sessions continued their practice of advertising for the construction of larger bridges and the *Exeter Flying Post* of 30 March 1809 contained the following advertisement:

NORTH DEVON
BRIDGE to be BUILT
Any person or persons willing to undertake to REBUILD NEW BRIDGE over the river Taw between South Molton and Barnstaple, nine miles from the former and three miles from the latter town are desired to send their proposals on or before the 21st day of April next, sealed up, to the Clerk of the Peace's Office, Exeter, where plans and specifications for the work to be performed may be seen.
March 24, 1809   R. Eales, Clerk of the Peace.

Presumably there were no satisfactory offers, for in the *Exeter Flying Post* of 27 April there was a repeat of the advertisement calling for tenders by 19 May. In May 1809 it was left to Green to contract to rebuild New Bridge over the river Taw for £3000, and Uton Bridge near Crediton for £1000. Green's work would have been superintended by the magistrates of the division, just as his repair work would have been. In the minutes of the Quarter Sessions there was repeated reference to repair work to individual bridges being carried out under the direction of the magistrates of the division for sums not exceeding an amount stated for each bridge. Contracts for repair work could also be let to tender by local builders.

The Quarter Sessions approved new rules for contracts at the Michaelmas Sessions of 1811, three years after Green's appointment. As proposed by Mr Taylor they were as follows:
1. Magistrates in each division shall have full power to contract for repairing and rebuilding county bridges according to plans and specifications of the County Surveyor as approved at Quarter Sessions.
2. The plans and specifications shall be sent to the magistrates who shall say where they can be inspected.
3. The magistrates shall appoint a day and place for receiving tenders (after two advertisements in newspapers and handbills in the county) 10 to 21 days after the advertisement. Magistrates to contract with the lowest bidder after receiving security for due performance.

Green's contract was continued for a further year and, obviously inspired by their surveyor, the magistrates established rules for widths and materials of construction for their bridges by class:
Class 1.   Bridges for turnpike roads only, of the most direct communication with the metropolis – 18 to 20 feet wide, tooled stone in ashlar work, string courses, coping and pilasters finished in the best manner.
Class 2   Bridges for other turnpike roads and roads much used – 15 to 18 feet clear, arches, quoins, string courses, copings and pilasters of tooled stone, the rest of rough masonry.
Class 3   Bridges for more private roads – 12 feet clear, finished in other respects like class 2.
No deviation whatever shall be made from the plans and specifications of the County Surveyor, either as to nature of materials, size of stone or workmanship, without an application to the magistrates – consent to be in writing on the back of the contract, the sum allowed in addition or the saving to be deducted. The magistrates have power to order piles or any other material not specified in the contract on receiving a report from the County Surveyor that the same are necessary to make the work secure.

At the same time the magistrates recommended to the court that at the next Epiphany Sessions when a fresh agreement would be made with the surveyor, that it should make an order that in future the surveyor should in no case be allowed to contract for rebuilding or repairing any county bridges. Green had obviously overspent on some of his works. So in January 1812 Green's contract from Michaelmas was confirmed at £400 per annum, but he was to survey the bridges twice a year and not to undertake repair or rebuilding work – with the exception of Cowley Bridge. Cowley Bridge was to be an exception because it was a particularly important bridge, the cost of which was to be shared with the city of Exeter. Green's estimate was £9000 and the bridge still stands as an ancient monument, across the rivers Creedy and Exe, carrying the road from Exeter to Barnstaple.

It is interesting to note that Mr Taylor's rules for contracts which were accepted at Michaelmas 1811 just preceded 52 Geo III c110 which received royal assent on 9 July 1812. This act had five clauses as follows:

1. Juries (i.e. Quarter Sessions) could contract for periods not exceeding 7 years for the repair of any bridge, gaol, prison or house of correction at a certain annual sum and Sessions should give public notice of their intention to contract, contracts should be entered in a book by the Clerk of the Peace and the most reasonable prices should be accepted. At Easter, Quarter Sessions could appoint two justices for any division to order immediate repairs by written order up to a cost of £20. The intention was expressed that this was to allow timely and immediate repairs to save great expenditure.

2. At any Sessions after such repairs the assembled justices could order any sum not exceeding £10 to be paid out of county rates although no presentment had been made, provided there was a certificate from two justices. Sessions had to be satisfied such repairs were reasonable.

3, 4. The term of appointment of the two justices expired each year after Easter.

5. Quarter Sessions could contract with the trustees of any turnpike, or with their surveyors or clerks, for the repair of roads over bridges to the limits on either side (300 feet) for terms not exceeding seven years even though no presentment had been made of insufficiency of the roads.

The year 1813 commenced with an order in the Epiphany minutes that Mr Green should agree with contractors for further repairs to the roads over the bridges and their approaches for the coming year. In the minutes there are records of payments by the treasurer for keeping the roads over the bridges in repair and these payments increased during the decade. One of these was for Emmett's (also known as Hood or Riverford) Bridge, which the Ashburton and Totnes Consolidated Turnpike had announced in the *Exeter Flying Post* of 27 June 1811 as being open for traffic. Though not recorded elsewhere, the County Surveyor must have proposed, and had accepted, this bridge as a county bridge.

The county was beginning to receive claims from landowners and so in the summer an order was made that the Surveyor must include in his estimate and specification all the works necessary to be done in repair or rebuilding and that the Surveyor should agree with the persons having property adjoining for land, fences, trees etc. as were necessary (under the superintendence of the magistrates of the division) before work was begun. The involvement of the magistrates ensured that these arrangements were fairly and honestly made.

At Midsummer 1814 Green made an application for an increase in salary citing his expenses in being away from home, income tax and the postage of letters. He also reminded the county that having completed the contract for Cowley Bridge and having a great quantity of implements, materials and men available he had asked to be allowed to contract for Teign Bridge and Chudleigh Bridge. Because in 1813 Green had overspent £1200 on a contract for Thorverton Bridge entered into as a result of an advertisement of 7 May 1811 in the *Exeter Flying Post*, the court refused to deviate from the rule laid down respecting contracts to repair and rebuild bridges, but he was offered a salary of

£500 per annum. After more representations, at Michaelmas 1814 he was allowed £550 per annum for a term of three years free and clear of property tax but it was confirmed that he was not to have anything to do directly or indirectly with the repair or building of any bridge. This agreement was extended to three years from Easter 1815.

At the Easter 1815 Sessions it was recommended that in future the securities for contractors for rebuilding bridges should be jointly and severally bound for the due execution of the works in one half only of the price contracted for. At Midsummer it was ordered that in future, copies of every contract and specification for building bridges be made and delivered to the contractor at the expense of the county.

The following year, at Michaelmas 1816, it was ordered that the present contracts for repairing the roads at the county bridges should cease at Easter 1817 and that the surveyor should give the necessary notices for the purpose and take the proper steps for reletting the roads; in the meantime he should estimate the expense of such repairs and deliver in the amount of the present contracts and consult with the magistrates of the several divisions respecting them. At the Epiphany 1817 Sessions it was ordered that the road repairs advertised to be let by tender should be let according to the plan delivered by Mr Green. Tenders would be returned to the Clerk of the Peace to be opened before a committee of the magistrates.

The way things could work in a hurry was shown at the Easter Sessions of 1817 with Weston Bridge was thrown onto the county by the judgement of the King's Bench. It was ordered that the surveyor prepare plans and estimates for repairing or building the said bridge immediately and report the same at the next adjourned Sessions when such directions respecting it would be given as the court might deem. One month later on 16 May, Mr Green having delivered a plan for building a new bridge it was referred to the magistrates of the division to examine the plan and estimate. They were to report their opinion at the next adjourned Sessions on 16 June when it was ordered that the bridge be rebuilt and the river diverted at an expense not exceeding £1400. At the Epiphany Sessions (13 January 1818) £600 was paid on account to Mr W.H. Lee, the contractor, and the balance due on the contract, £200, was paid on 31 March 1818 at the Easter Sessions. Weston Bridge, sometimes known as Trafalgar Bridge, is a three-arched bridge of main span 25 feet, just west of Honiton at map reference ST143001.

The year 1818 also saw James Green taking a rôle in county buildings in Exeter. The main county buildings were the castle, lying within Exeter city walls and three buildings without, the gaol, the Bridewell and the sheriff's ward. The gaol had been built under an Act passed in 1787 and was located north of the castle on the far side of the Longbrook. Alongside the gaol was the Bridewell, later called the house of correction, which stood on the south-east side of St Thomas (now Cowick) Street in 1805, where it can be identified on a plan of Exeter. The Bridewell was removed to the site alongside the gaol in 1810. The gaol normally accepted prisoners who were convicted for execution or for transportation

BRIDGES, LAND RECLAMATION AND ARCHITECTURE

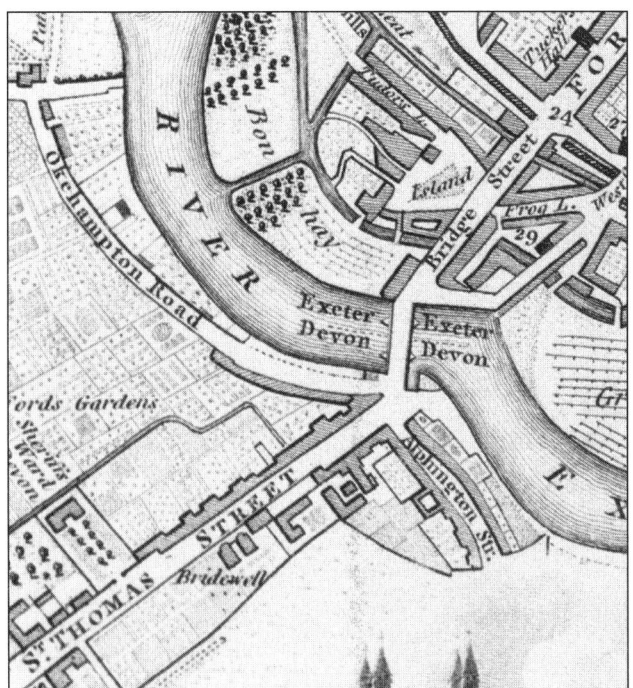

Extract from map of Exeter, 1805, showing St Thomas Street

whereas the Bridewell housed persons who had been given terms of imprisonment.

The sheriff's ward was the debtors' prison and it was located on the north-western side of St Thomas Street further out of Exeter than the Bridewell had been. All that is left of this building today is a stone gateway at the junction of Cowick Street and Buller Road.

At the Epiphany Sessions of January 1818, the sheriff's ward committee was empowered to call in two surveyors to discuss whether the present ward could be repaired and made secure or whether to advertise for plans for a new ward. At the next Sessions on 31 March the committee recommended the plans and specification of Messrs Cornish and Son for building a new ward at a cost of £8000. It was recommended that a contract be made and work carried out under the superintendence of Mr Green, the County Surveyor, the buildings to be completed by 1 November 1819.

It was also ordered that Mr Green continue as bridge surveyor and surveyor of all the county buildings and land at a salary of £550 per annum for three years from this date. This period ran on directly from his previous £550 per annum contract of Easter 1815, with no increase for the extension of his responsibilities. The title of County Surveyor was now more appropriate as describing his work. At the Easter Sessions his extension of responsibility and pay were confirmed to extend for three years from that date. Messrs Cornish and Son's tender of £8000 for building the new sheriff's ward was formally accepted and a contract prepared, the work to be under the superintendence of Mr Green. He was also ordered to put the castle in proper and complete repair at as reasonable expense as possible. One can only contrast the care of the Sessions records about the ward contract and Green's functions with the almost casual record of his work to be undertaken at the castle with no financial limit.

In 1819 an application by Green to be reimbursed for expenditure on tolls of horses and carriages employed in the building of Cowley Bridge was successful and he was paid £50 4s 4d. Then other contractors applied for reimbursement of tolls paid for carrying materials for other bridges (Mr Lee of Weston Bridge made an application) but the magistrates decided to investigate the toll keepers' accounts instead. Eventually in July 1819 the chairman was requested to write to the county members to bring in a Bill at the next session of Parliament to exempt all materials for building and repairing bridges and the roads over them from tolls in the same manner as materials for highways were exempted by the Act of 13 Geo III c84 s60.

At the Easter Sessions of 1820 the treasurer and bridge surveyor were ordered to deliver at all future Sessions their reports in writing to the chairman on his arrival at the court on the first day. Such reports would be the first subject on the county business to be taken into consideration by the magistrates on the adjournment of the court on the first day.

At Midsummer it was ordered that when money not exceeding a specified sum had been granted at any Quarter Sessions for the execution of any county work, and an authority given to the justices, acting in the division wherein such work was situated, to contract with the person or persons making the lowest tender for the completion of such work, the justices so contracting should first transmit the contract to the clerk. Then the justices would have full powers to make such orders on the treasurer for payment by such instalments as they thought fit. Also the instalments should be paid by the treasurer on demand provided that such orders did not exceed the sum originally demanded and the County Surveyor, by his signature, had certified that such payments should be made.

The procedure was now clearly devolved. At Quarter Sessions an order would be made for a bridge to be rebuilt (or repaired) for a specified maximum sum. The magistrates of the division would be responsible for obtaining tenders, they would accept the lowest tender, and decide the instalments by which it would be paid. With the contract sent to the clerk, the treasurer would be bound to pay on demand by the local magistrates the instalments as they were certified due by the surveyor and referred by the magistrates to the treasurer.

Towards the end of 1820, William McAdam, the surveyor to the Exeter Turnpike Trust, agreed to take over the maintenance of the county bridge roads (over the county bridge and 100 yards in either direction from the abutments). It was agreed that he would be at liberty to expend a sum not exceeding £90 per mile for the first year and £40 per mile for each succeeding year and would be paid a fee of £25 per annum for his supervision.

So by the end of 1820 the County Surveyor was responsible for preparing plans and estimates for county work and had the responsibility of certifying the payments that were due to contractors. Above all he was responsible for the bi-annual examination of the bridges.

2.22 BRIDGES RENEWED AND REPAIRED 1809–1820

Before 1773 it was the law that highways to a market town should not be less than 8 feet wide and that no horseway should be less than 3 feet wide. Local turnpike acts show that larger ideas were prevailing and by the Exeter Act of 1753 power was given to widen roads to a width of 30 feet. Devon was not a rich county and the bridges were therefore narrow between the parapets and composed of a few small-span arches. In times of flood, water might be expected to inundate the road approaches to the bridge. Not surprisingly, Quarter Sessions was charged with maintaining the 100-yard stretch of road on either side of a county bridge.

From medieval times bridges in the south-west peninsula had been built with spans of 15 to 20 feet. The larger the river the more spans were built, Staverton Bridge over the river Dart near Totnes and the medieval Exeter Bridge over the Exe are fine examples of this principle. In the late eighteenth century, Exeter Turnpike Trust built Countess Wear Bridge with seven spans which ranged from 23 to 26 feet while the medieval Exe Bridge was replaced in 1778 by the Trust with a bridge composed of three large spans of $37^1/_2$, 50 and $37^1/_2$ feet each. These were exceptions to the general rule in Devon.

Having been appointed on 4 October 1808, Green set about compiling a report on the condition of the county bridges that he would have to maintain. This report was completed and presented three months later to the Quarter Sessions (Epiphany) on 10 January 1809.[2] Noting the spans and the widths between the parapets of each bridge, Green gave a description of the materials of construction and the condition of the structure.

An examination of his reports shows that only nine of the 236 bridges had a span or spans of 30 feet or more. The nine bridges were Exe Bridge, Morebath, with spans of 25, 30 and 25 feet; Fenny Bridges which he had just rebuilt with spans of $41^1/_2$, 47 and $41^1/_2$ feet; Lee Mill Bridge, also just rebuilt with a span of 36 feet; Last Bridge, Cullompton, 30 feet; Pynes Bridge, Upton Pyne, 30, 16 and $9^1/_2$ feet; Thorverton Bridge (1796) over the Exe, 30, 40 and 30 feet; New Bridge, Tawstock, then about to be rebuilt, at 30, 40 and 30 feet; Tarr Bridge over the river Yealm spanning 30 feet, and Gara Bridge, Diptford and Halwell, 30 feet.

27 bridges had a width between parapets greater than 12 feet, but 110 had a width of only 9 feet or less. Devon appeared still to be in the pack-horse age. In contrast Green was careful to build substantial structures with large spans commensurate with the often large catchment areas of the rivers

Drakeford Bridge 1809 (SX789801)

Yeoton Bridge 1809   *(SX827988)*

passing beneath. Green must have recognised that large spans mitigated the effects of scour on piers and abutments in time of extreme flood, when the steep slope of so many Devon rivers raised water velocities to exceptional levels.

Within a few years many important bridges had been built. Fenny Bridges and the bridge over the river Yealm at Lea Mill were completed together with two single spans, Glazebrook for the Exeter–Plymouth road at South Brent and Uton (Yeoton) over the Yeo south west of Crediton on the road from Crediton to Tedburn St Mary. In addition to this last bridge, Green contracted to build New Bridge over the river Taw at Tawstock, a graceful structure for this important river.

Another important structure was begun at about this time. We have noted that in July 1806 Green was instructing a surveyor on behalf of the Duke of Somerset at Totnes. An act had been passed in 1805[3] for a new turnpike road from Totnes to Ashburton to follow the line of the river Dart and to cross this main river at a place called Emmett's. The Ordnance Survey map of 1809 showed the road in place but not the bridge, but in 1811 the *Exeter Flying Post* carried an advertisement announcing that Emmett's Bridge was open 'for cattle, carts and carriages'. Furthermore the Epiphany Quarter Sessions minutes, January 1813 noted that Thomas Brown had been paid £17 19s for repairing the roads over Emmett's Bridge. It must therefore have been a county bridge and to achieve this status would have had to be approved as satisfactory by the local magistrates on the advice of the County Bridge Surveyor. An examination of the bridge shows that it was 18 feet wide between parapets with spans approximating to 35, 40 and 35 feet and very much in the style of Green's three-span bridges. Green probably used his local knowledge to persuade the Totnes turnpike authorities to pay him for supplying the drawings and specification before he supervised the construction of the bridge, though it must be noted that there is no factual evidence of this.

A priority list for the Easter Quarter Sessions of 1809 had identified Cadhay Bridge, Hele Bridge, Hatherleigh, and Thorverton Bridge for rebuilding. Cadhay, over the river Otter near Ottery St Mary and Cadhay House, needed rebuilding for previous work had not lasted long. (Recently repairs disclosed a stone bearing the date 1760.) The advertisement for the reconstruction of Hele Bridge, Hatherleigh was carried in the *Exeter Flying*

Emmett's Bridge (also known as Riverford or Hood Bridge) 1810–11 (SX772637)

Post in May 1810 and that for Thorverton in May 1811.

Meanwhile at Epiphany 1810 another New Bridge, this time between Chudleigh Knighton and Kingsteignton was ordered to be built at a cost not exceeding £500. This replaced a 9 feet wide bridge with two 16 feet spans and one 10 feet 6 inches span over the river Teign. Now, after further attention, the bridge has spans of 20, 22 and 12 feet but its width between parapets is only 10 feet 6 inches. This modestly constructed bridge was far outranked by Hele Bridge, Hatherleigh with its spans of 34, 38 and 34 feet over the river Torridge.

When Thorverton Bridge was ordered to be rebuilt at Easter 1811 there must have been a lack of applicants to carry out the work because in contradic-

Cadhay Bridge 1809   (SY093960)

Hele Bridge 1810–12   (SS540063)

tion to Quarter Sessions policy Green was awarded the contract. He constructed a massive arch of 84 feet span with a rise of 26 feet with subsidiary arch spans on the approaches. The eastern arch was approved at an additional cost of £650 at the Michaelmas 1813 Sessions when, having represented to the court that 'he had lost £1200 in consequence of having been disappointed of stone from a quarry he calculated to have got therefrom at the time he entered into the contract', Green was allowed only £600.

Such was the height of the arch that extra horses had to be used to move loads over the bridge and such was the pressure on the spandrel walls that they began to fall in 1906 so that in 1908 the main and subsidiary spans were replaced by a reinforced concrete arch bridge at a lower level so close by the former bridge that the former abutments remain as river training walls on the river approach to the replacement. The 84 feet span at the lower level is quite satisfactory for river flows today.

When Thorverton Bridge was approved, three new small bridges were ordered south of Newton Abbot over Aller brook at Keyberry, Langford and Aller. The advertisement for these three bridges and for New Bridge, Kingsteignton, appeared in the *Exeter Flying Post* on 27 June 1811.

> Persons to rebuild these bridges to send proposals 6 July next, sealed up, to the Clerk of the Peace at Castle, Exeter, where plans and specifications can be seen. On 9 July Magistrates of the division will attend the Globe Inn, Newton at 12 o/clock to take proposals into consideration.

In 1812 the county magistrates were persuaded to repair Membury and Cotleigh Bridges on the Chard turnpike road to Honiton of 1776 which had severe gradients to cross the rivers Yarty and Corry. By a change of county boundary, both bridges still carry a Dorset notice that threatens transportation to

BRIDGES, LAND RECLAMATION AND ARCHITECTURE

Thorverton Bridge 1811–13: this picture shows the 1811–13 bridge in the background with the 1908 bridge under construction in the foreground. (Photo by courtesy of D.I. Stoyle) (SS936017)

anyone damaging their structure! Another bridge near Dorset's boundary built in 1811 was Uplyme Bridge.

Cowley Bridge had collapsed in a flood on 25 January 1809 but its reconstruction was delayed for some years as it spanned the boundary between the borough of Exeter and the county of Devon. At Easter 1813 the magistrates of Devon and of Exeter requested the County Surveyor to meet the city surveyor to draw up plans and estimates for reconstruction. At Midsummer the court ordered that a contract be entered into with Mr Green for rebuilding the bridge in the sum of £9000. Work seems to have proceeded rapidly and by 9 June 1814 Green was advertising for 'ten or fifteen Prime Mallet and Tool Hands who have been also used to fitting fine and heavy ashlar masonry.' On 6 October 1814 the *Exeter Flying Post* had 'great pleasure in announcing that the new Cowley Bridge being finished it was

Long Bridge, Membury 1810–11   (ST255055)

on Wednesday 5 October opened for the public service after an interval of nearly six years ... '.

Green produced a fine bridge of three spans of 50, 55 and 50 feet which is now scheduled as ancient monument, no. 163, in the county list. It carries water from the Creedy and part of the flow from the Exe and Culm catchments and is typical of Green's careful provision for the flow from the occasional extreme flood.

Cowley Bridge bears a remarkable resemblance to Wolseley Bridge over the river Trent, north-east of Rugeley, in South Staffordshire (SK 021 204). This bridge was built in 1799 to the design of John Rennie at a time when Green was on Rennie's staff, so one can imagine that Green used his association with Rennie to repeat a splendid structure over an important river.

Cowley Bridge 1813–14   (SX907055)

Green had been allowed to contract for this bridge because he was producing a particularly fine bridge for both the city and the county. But despite his pleas the court still would not permit him to contract for county bridges although he used all his powers of persuasion to be allowed to use his equipment for Chudleigh and Teign Bridges. An advertisement in the *Exeter Flying Post*, 10 November 1814, put up for sale 'several very good and tried cart horses, all of whom may been seen at their work at Cowley Bridge near Exeter.' Green's fine bridge still stands, carrying the important road from Exeter to Crediton and thence to Barnstaple. It is now scheduled as an ancient monument and listed as a building of historical interest, Grade II.

In 1813 the road from Exeter to Barnstaple proceeded from North Street, Exeter via St David's

BRIDGES, LAND RECLAMATION AND ARCHITECTURE

Head Bridge 1813   (SS667182)

Hill, Cowley Bridge, Crediton, Morchard Bishop, and Chawleigh to Chulmleigh. Here there was a choice of routes, one of which crossed the river Mole at Head Bridge and continued via Chittlehamholt to Barnstaple. The other route descended to the river Taw at Colleton Mills, crossed Hensford Bridge and continued via Burrington and High Bickington to New Bridge Tawstock and on to Barnstaple.

Head Bridge on the river Mole was rebuilt in 1813. Colleton Mills Bridge over the river Taw was widened between 1815 and 1817 for highway traffic, emphasising the importance of that route to Barnstaple as an alternative to the route over the river Mole. £300 was allowed in 1815 and a further £60 in 1817.

In October 1813, Mr Aust, the Surveyor of the Post Office, informed the court that Chudleigh Bridge was too narrow and dangerous for public carriages to pass and that the Government had given orders for the inspector of mail coaches to indict the court. This produced a flurry of activity with Green being ordered to draw up a scheme for the necessary improvement. So by 1814 Green was producing drawings and specifications for this bridge and also for another bridge over the same river, Teign Bridge. Chudleigh Bridge was reconstructed to a span of 60 feet in granite and when some alterations were made in 1973, it was necessary to go to Merrivale quarry on the other side of Dartmoor to match the stone.

Before 1814, Teign Bridge consisted not only of a main bridge over the river channel of two 15 feet spans, with 11 feet between parapets, but was part of a causeway with subsidiary arches beneath, running south-west to Newton Bushel and Newton Abbot across a plain now extensively worked for china clay.

Chudleigh Bridge 1815   (SX857784)

31

New Bridge, Tawstock 1809 (SS570280)

Bridges, Land Reclamation and Architecture

P.J. Taylor, the local magistrate, recorded the reconstruction in detail.[5] Curiously enough the 1814 bridge was built above a five-span arched bridge, possibly thirteenth century, which had either sunk very slowly or the ground had risen, for the springings of these old arches were at low water level. The bridge to be replaced had been built above the first two arches. Green used the first and the third openings of the older bridge to excavate for his foundations which were 15 feet 5 inches below the waterline and 7 feet further below the meadow. At this level the foundations of Green's bridge were below the level of the earlier bridge foundations.

Green's new Teign Bridge was completed with a 50 feet single span and a road width commensurate with his guidelines. There are no signs of settlement but Gregory Weatherdon, the contractor, was allowed a further £660 in 1817 for losses incurred because of additional works.

In 1765 there was a route across the river Torridge at Sheepwash which ran north through Marland and Monkleigh to Bideford, across Landcross Mill Bridge over a tributary to the river Torridge. This single arch bridge was rebuilt in 1815 but nowadays the road it carries is little more than a farm road as the route was diverted only a few years later.

Meanwhile in 1814, Ash Bridge, Throwleigh, Pord's Bridge, Stoke Gabriel and Steps Bridge, Dunsford had been ordered to be rebuilt. The first two were modest single-span bridges, but Steps Bridge, completed in 1816, was another substantial three spans over the higher part of the river Teign. Here the contractor deviated from the contract and for some time money was held back, the final payment not being made until Easter 1818, although it had been contracted for £1970 in January 1815.

Teign Bridge 1815–16 (SX859734)

Landcross Mill Bridge 1815 (SS454235)

Steps Bridge, Dunsford 1816 (SX804883)

From time to time problems arose due to other authorities wanting the court to take over the maintenance of their new bridge. In January 1812, a bridge over the Withycombe brook under the Exeter to Exmouth road, and a bridge in the Dawlish Strand were referred to the magistrates to ensure that the parishes built them to statute. We hear no more of these but at Sessions in July 1813 Green produced a note that he had received from the surveyor of the parish of Hemyock requiring him, as surveyor of the county bridges, to superintend a bridge about to be built over the river Culm. He was ordered to report at the next Sessions which he did, saying that it had not been built as to be approved. The court decided that in such circumstances they would not be bound to repair the bridge in future. In October 1815, Yarcombe Bridge, neither having been built under the inspection nor to the satisfaction of the County Surveyor, the court ordered that notice be immediately given to Mr Bowden, the clerk of the trustees of the Chard turnpike road that the county of Devon would not consider themselves liable to take that part of the bridge which was in the county as one of their county bridges. Furthermore, Green was ordered to make a written report on this to be filed among the records of the county. On this occasion Green appears to have over-reacted for this bridge survived the catastrophic flood down the river Yarty in 1968.

In 1817 Bunker's Bridge, Chudleigh Knighton, carrying the Exeter to Plymouth road was repaired by Thomas Brown while Weston Bridge over the river Otter near Honiton was rebuilt as a three-span structure. Money then had to be spent to maintain the course of the river on its approach to the new bridge, possibly because its northern tributary, the river Wolf, had been diverted upstream away from the bridge site. Stoke Canon Bridge was also receiving attention, John Perryman being paid £140 in October, with £200 allowed in 1818 and Perryman being paid a further £109 for widening the arches in October 1819. This bridge over the river Culm, near its confluence with the river Exe, has a long causeway with subsidiary arches beneath.

Nicholas Robins had to apply to the Sessions for extra payment for widening Rushford Bridge, Chagford, in 1818 while Fairmile Bridge on the

Weston (Trafalgar) Bridge 1817 (ST143001)

Exeter to Honiton road was widened at a cost of £200 by Messrs Lee in 1819 and Downes Bridge on the Exeter to Crediton road widened by Mark Helmore for £145. John Perryman widened Withy Bridge on the Exeter to Broadclyst road for £80 in 1820. These improvement works on modest bridges showed that, with increasing traffic, people were no longer content with the old single file working. Other bridges widened at this time were Shaugh Bridge over the river Plym and Topsham Bridge over the river Avon north of Loddiswell.

In 1820 the Sessions ordered Last Bridge, Cullompton, to be rebuilt at a cost not exceeding £400. This bridge took a road from Cullompton over the river Culm south-east of the town and leading to Plymtree. This is quite a large single-span bridge and a similar bridge was ordered at

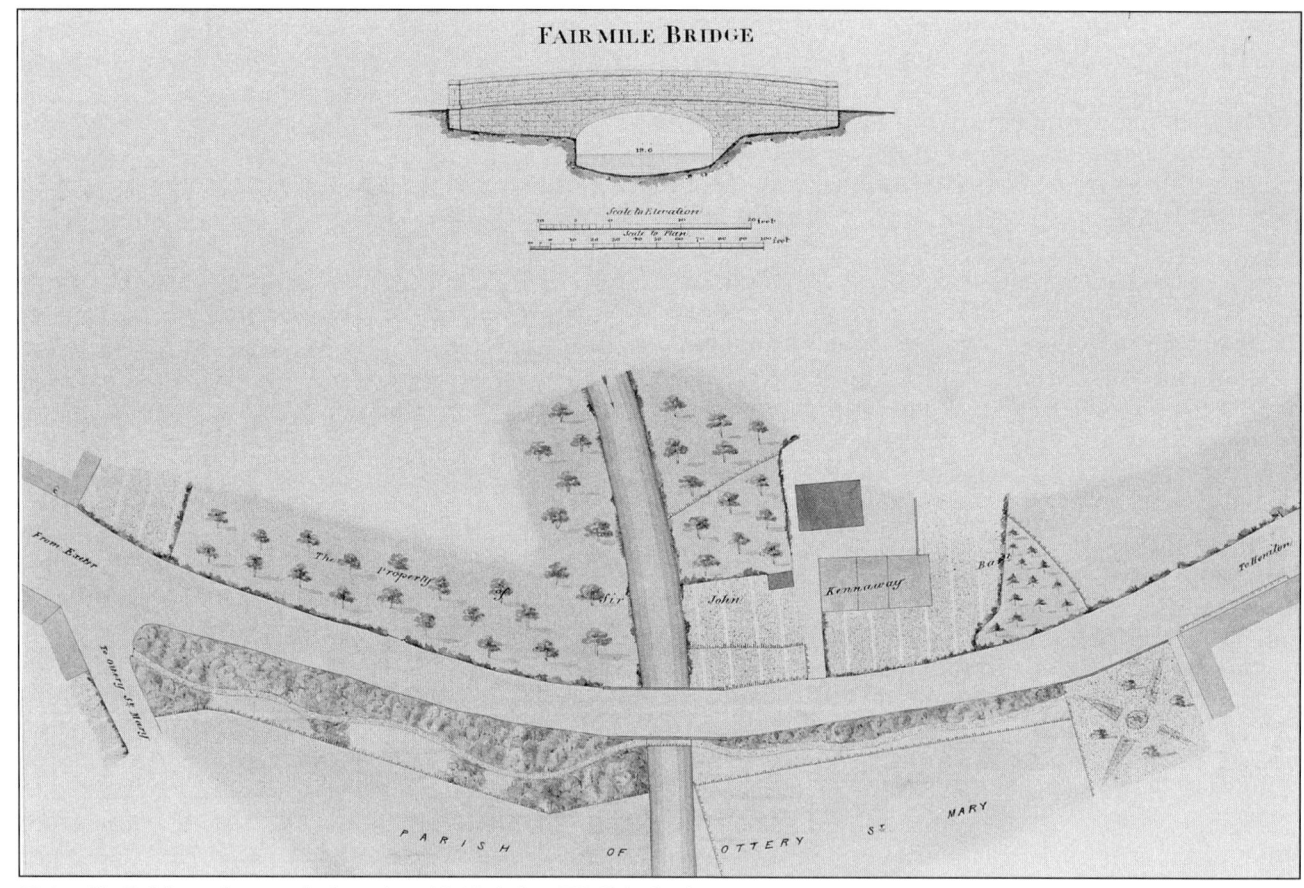

Fairmile Bridge, plan and elevation 1818 (after T.Whitaker) (SY088971)

Dogmarsh, on the Moretonhampstead to Whiddon Down road to carry it over the river Teign. These powerful rivers required sturdy bridges and the latter is also very attractive.

Perhaps the last venture for 1820 was the most pleasant. Mr Savile, a member of the court, having stated that Brightley Bridge in the parish of Okehampton stood in a very 'ineligible' situation, offered, at his own expense, to build a new bridge in its stead on a more commodious spot. It was ordered that his offer be accepted provided such spot should be approved by the magistrates of the division and the County Surveyor and that such bridge be erected under their supervision.[6] So no

BRIDGES, LAND RECLAMATION AND ARCHITECTURE

Last Bridge 1820   (ST027066)

doubt another drawing and specification from Green was required.

2.3 BRIDGE WORK FOR OTHER AUTHORITIES

2.31 BIDEFORD LONG BRIDGE

At the turn of the nineteenth century many bridges were being widened to improve traffic flow. Barnstaple Bridge was widened in 1796 and widening began in 1795 on Bideford Bridge and was about half finished in 1810 when the trustees gave orders to Green, as the contractor, to complete the widening of the rest of the bridge and at the same time rebuild the parapets in Penarth ashlar stone.[7] The widening was carried out by springing segmental arches about 2 or 3 feet wide from the cutwaters as was the usual practice in those days.

On 3 August 1810, Green made an agreement with the Trustees to widen the nine eastern arches in blue lias ashlar stone for £2130.[8] He also had to take down four cutwaters on the south side of the bridge and one on the north side and extend them

Brightley Bridge 1820   (SX598974)

to enable the parapets to be built in a straight line as the segmental arches required all the cutwaters for support.

Whether Green made a profit on this contract we do not know. Certainly it was binding on him and his executors to complete the work. We have noted that in 1811 the Quarter Sessions forbad Green to enter into any more contracts with them because he had overspent. Green was unusual among civil engineers in combining county work, consulting work and contractual work and one fears that his fortunes may have swung wildly from profit to loss for there was no limited company in those days.

2.32 THE RIVER BRIDGES AT EXETER

The bridges at Countess Wear and at Exe Bridge from the city to St Thomas were not county bridges

37

Bideford Bridge as shown on a print of 1835 (SS456263)

but had been erected by the Exeter Turnpike Trust. Thomas Parker of Topsham built Countess Wear Bridge on behalf of the trustees. It was opened on 14 September 1774 with arches ranging in size from 23 to 26 feet span. In October 1842 one of the two middle piers was removed and the central arch and its neighbour replaced by an arch of 60 feet span to a plan by Thomas Whitaker, who was by then about to become County Surveyor in succession to Green.

The city bridge, which replaced a bridge that had been built in 1773 and washed away in 1775, was begun in 1776 and completed in 1778 by Joseph Goodwin. It was a three-arched structure with spans of 37½, 50 and 37½ feet and by the end of 1810 concern was being expressed about its condition to withstand floods. James Rowe was the surveyor to the Trust and he embarked on work in 1811 to remove projections into the river on the inside of the bend upstream of the bridge (the north-eastern side). This was causing the river to accentuate its normal tendency to deepen on the outside of any bend, in this case on the south-western side of the channel as it approached the bridge.

James Green was asked by the trustees to report on the work Rowe was carrying out and to make his own recommendations which he did on 2 June 1811.[9] These were as follows:

    a. On the north-east side of the river no point of land ought to be left projecting into the river beyond the line of the abutment for about 200 feet upstream of the bridge and then the side of the river should be gradually curved from the upper end of the Bonhay. Excavation to give this shape should be sunk as deep as possible below the common low water of the river.

    b. A part of the south-west side which receded beyond a parallel line with the north eastern side distant the span of the arches and the width of the piers should be filled to resist the force of the water and extend until it meets the natural curve of the river.

    c. The bed of the river for at least 300 feet upstream of the bridge should be as nearly as possible level and have a graded and regular descent to the bridge.

These recommendations were accepted and despite subsequent floods and the rather narrow waterway the bridge stood until it was replaced by a clear span of 150 feet in 1905.

2.4 CREDITON AND TORRINGTON CANALS

A canal between Exeter and Crediton was first proposed in 1792 and the line was surveyed by Robert Cartwright in 1800 and 1801. James Green advertised for canal cutters in the *Exeter Flying Post* of 1811 but little is known thereafter.[10] While excavating for the widening of the river Exe in May 1973, the remains of a lock of the unfinished scheme, some 80 feet long and 14 feet wide, were found but nowhere were they more than 7 courses high. The remains of a 16 inch diameter brickwork pipe were also found..

In 1810, 14 years before the construction of the Torrington Canal, James Green carried out a survey and prepared a plan for a canal on the eastern side of the river Torridge. It extended from Torrington south past Wear Gifford to join the river just above Hallspill and opposite Landcross. A plan was deposited but nothing was done until the scheme on the western side of the river was commenced in 1823.

2.5 LAND RECLAMATION

Among the noblemen solicited by James Green in his advertisements in the *Exeter Flying Post* of 1808 was Lord Rolle, owner of extensive lands in North and East Devon. Having successfully enclosed Chelson Meadow and land at West Charleton for

Lord Boringdon, it was natural that Green should propose similar works for Lord Rolle and that when these were successfully completed Lord Rolle commission Green to build the Torrington Canal.

2.51 BRAUNTON MARSH

At the time of the first Ordnance Survey of North Devon in 1809, Braunton Marsh was shown as tidal land on the Taw estuary but the Burrows and the Great Field were not. At Braunton on 14 September 1809, Green proposed a scheme for Lord Rolle and other landowners whereby he would construct for under £20,000 an embankment 4000 yards long from the burrows alongside the Braunton Pill to Velator Bridge where the river Caen joined the Wrafton brook.[11]

A survey of the Marsh by Green and John Pascoe led to the deposit of a plan in September 1810 with the Clerk of the Peace[12] and to the Braunton Land Enclosure Act of 1811, with works being completed by about 1814. As at Chelson Meadow, the embankment sloped at approximately 1:4 on the estuarial side and at 1:1½ on the landward side, with a crest 6 feet wide. A roadway and a boundary drain were placed on the land side of the embankment. A tidal outlet was built at SS477342 and this now discharges into Horsey Island, an area of land subsequently enclosed by N. Whitley in 1857. Approximately 1300 acres were enclosed by Green and 210 acres by Whitley. This extensive grazing area is now administered by the Braunton Land Drainage Commissioners. Subsequently, drainage channels were constructed by a contract dated 10 September 1814 between Green, on the one hand, and Lord Rolle, with Messrs Cleveland, Drake, Dyer and Webber, on the other.[13]

Braunton Marsh Embankment (SS469327 – 485357)

At the northern end of the embankment the river Caen was raised above normal tide limits where it passed beneath Velator Bridge (SS486357) to discharge into the tidal river.

In September 1821 Green put forward another plan showing a proposal for a canal from the Taw estuary at Broad Pill to the north sluice at Braunton embankment, that is, following the line of the river Caen where it is tidal.[14] No more was done, however, in his lifetime. John Pascoe subsequently (about 1825) became a surveyor to the Exeter Turnpike Trust being engaged in route location for new works.

2.52 BUDLEIGH SALTERTON

The wide low valley of the river Otter in East Devon is almost closed at the sea by a shingle bank and for 2000 yards inland the 500 yard wide estuary was subject to inundation at high tides and storms until Lord Rolle commissioned Green to reclaim an area approximately 2000 yards long and 330 yards wide, thereby enclosing over 140 acres.[15, 16]

The original course of the brook through East Budleigh turned due south to run below the low cliffs on the west side of the estuary. The Budleigh brook was diverted on to an aqueduct to discharge into the river Otter upstream of these works. Ditches across the reclaimed land drained to the original course of the brook below the sandstone cliffs, and the water outfalls to the sea in a culvert below the shingle bank. Construction appears to have been carried out in 1812–13, possibly with the assistance of French prisoners-of-war.

2.53 JOHN RENNIE AND THE ROYAL NAVY

Green's association with John Rennie did not end when he became County Surveyor. Apart from the Killerton water supply proposal (see 2.74), another occasion arose from Rennie's concern that the tidal flow into and out of Portsmouth harbour would be impaired by any land reclamation in the harbour and that the liability to silting of the naval dockyard would thereby be increased.

There is a large tidal inlet almost reaching the church in the parish of Alverstoke on the Gosport side of the entrance to the harbour. Apparently the trustees of the parish had made a causeway across this Stoke lake to the poor house preventing water from flowing into and out of some 14 acres of land at high water. The agent to the solicitor of the Admiralty therefore wrote to the Reverend Richard Bingham and the trustees of the parish on 4 October 1813 giving them a month to remove the causeway or His Majesty's attorney general would file information against him in the Court of Exchequer to have the same abated and removed.

The Admiralty solicitor sought professional advice to back up his request and the following statement was produced on 14 October:[17]

> We the undersigned having at the request of Mr Grantham the Agent of the Solicitor of the Admiralty at Portsmouth inspected several encroachments stated in the report of the respective officers of His Majesty's Dockyard there dated 22 January 1813 are of the opinion that the causeway made across an arm of Stoke lake forming a road to the Poor House of the parish of Alverstoke is highly injurious to the Harbour of Portsmouth by preventing a very considerable quantity of water from flowing over nearly 14 acres of land above the same on every influx of the tide and is a great nuisance to the Harbour as are also the erection carried below the High Water Mark at the back of the Quebec Tavern occupied by Mrs Stevens, the Wharf erected on the Camber by Messrs Burridge and Sons at the back of their Storehouse, and the embankments made by Mr Parfitt, Mr Geo. Adrian and Mr James Vaughan in Stoke lake – and that it is advisable to cause the

same to be abated by immediate proceedings under the direction of His Majesty's Attorney General, and the other minor encroachments therein stated to be removed.
RALPH WALKER   J. WHIDBEY   JOHN RENNIE
JAMES GREEN
WILLIAM CHADWELL MYLNE   JOS. JESSOP
WILLIAM CHAPMAN

A plea was made on behalf of the trustees of the parish by George Porter, Lt. General, Government House, Gosport, in his capacity as magistrate and as guardian and visitor for the poor. This brought a conciliatory reply from the Navy Commissioner insisting on action but suggesting an opening in the causeway 80 to 100 feet long using 12 feet span brick arches or wooden piles.

So within five years James Green had changed his views from the enclosure of Chelson Meadow in the Plym estuary to ensuring that inlets were maintained to allow the ebb and flow of tidal water. Over 30 years later Green would give evidence on behalf of Exeter City restricting Brunel from blocking inlets in the Exe estuary when the South Devon railway was promoted in Parliament, so that at least after 1813 his view was consistent and in 1813 he had agreed with some of the most eminent engineers alive. Against this one has to remember that he was at that time reclaiming Braunton Marsh for Lord Rolle and others!

2.6 ELMFIELD HOUSE

Soon after Green arrived in Exeter he set about building himself a home and the one he produced was particularly impressive even for a young professional man. An extract from the Order Books of the Exeter Turnpike Trust of 1 March 1814 described 'part of a road descending from St David's hill as fronts the house occupied by Mr James Green' and in the *Exeter Pocket Book* of 1816 Green's address was given as St David's Hill, while in 1822 it was Elmfield, St David's Hill. Yet by 20 April 1826 an advertisement in the *Exeter Flying Post* announced that Elmfield and its estate was for sale and that it had lately belonged to George Sparkes Esq.

Elmfield House is shown quite clearly on the OS second edition 1905 1/2500 plan LXXX.6 originally surveyed between 1875 and 1888. There is also a plan of Elmfield and its grounds in a document of 11 October 1832 where the house is described as Elmfield House and Offices.[18] With this plan is an affidavit made by two gardeners testifying to the ownership of a piece of land in the north east corner of the plan. The plan gives the owner as George Sparkes. In the affidavit the gardeners mention that the property was formerly owned by James Green. The plan is signed in the bottom right corner by Samuel Coleridge and one assumes therefore that the ownership of a piece of land required by the Exeter Turnpike Trust for the New North Road was being established. At some time, therefore, between 1822 and 1826 Green sold Elmfield to George Sparkes assuming Green to be both owner and occupier.

This property, for a time the Imperial Hotel, has now been converted into a public house called The Imperial but Green's architecture is impressive. J.S. Rowe has described it in the following terms:

BRIDGES, LAND RECLAMATION AND ARCHITECTURE

Elmfield, St David's Hill, now The Imperial

James Green's House, Exeter (now part of the Imperial Hotel). Erected c1810 (George III) and having a three bay frontage of ashlar stonework probably from the Heavitree Quarry, with a simple cornice moulding, dentil and parapet. A projecting central entrance portico has two pairs of Tuscan style columns with Ionic capitals supporting an entablature with a plain frieze and cornice. The house has been much extended at various periods in red brick with slated roofs.

Before New North Road was built, the property was entered from the corner of Howell Road and St David's Hill where there is an original lodge. From the house the ground slopes steeply towards St David's Hill (and the station) and there are some fine trees in the grounds which extend to about 4 acres. As built, the house would have had agricultural land to the east, but the construction of New North Road changed that situation.

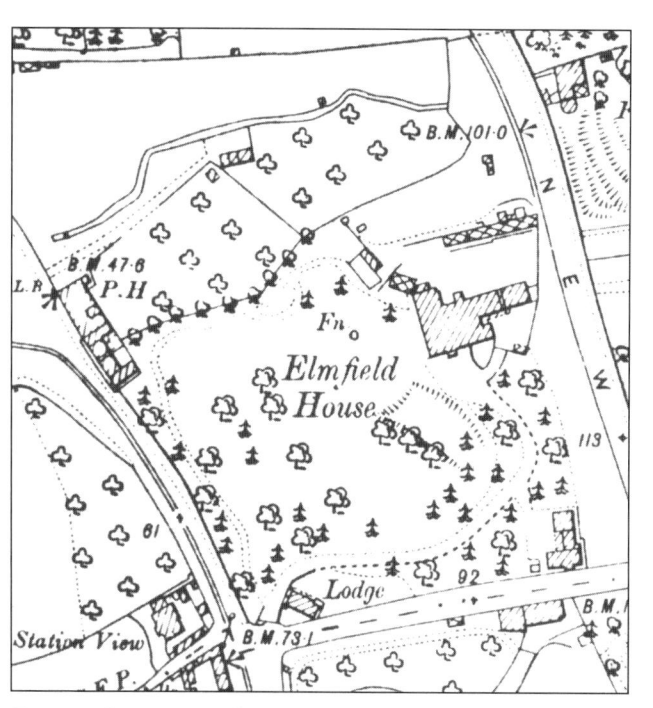

Extract from an early OS map showing Elmfield House and grounds

On 22 March 1826 an important Act was obtained by the Exeter Turnpike Trustees which enabled them to proceed with many schemes, one of which was 'A new road from the New London Inn in Exeter by the Eastern side of Elm Field House, to Duryard Lodge Gate into the road to Cowley Bridge.'[19] On 7 August 1833 the line of the New North Entrance was approved at a cost of £7500 and on 28 August 1834 the New North Entrance from Belmont through the Gaol Fields to the New London Inn was opened.

43

The name Elmfield House may well have been derived from the field in which it was built. The following advertisement appeared in the *Exeter Flying Post* on 19 January 1809.

> To be sold, 34 Elm Trees of large dimensions now growing in a field on St David's Hill. Auction at the Plume of Feathers, Saturday 28 January. For viewing apply Messrs Hicks and Son.

2.7 OTHER ARCHITECTURAL WORKS

In his early years in Devon, Green undertook a number of important architectural commissions. In 1810 he was working at Buckland House, and in 1816 rebuilding St David's church, almost on his doorstep in Exeter. Taking into account Elmfield House, Green showed himself to be an ingenious and competent builder of large houses who naturally earned the title of architect as well as civil engineer.

2.71 BUCKLAND HOUSE, BUCKLAND FILLEIGH

In 1810 Green undertook an important architectural assignment for John Inglett Fortescue when he transformed Buckland House which had been damaged by fire in 1798. Cherry and Pevsner say his work here showed him to be an accomplished innovative practitioner in the neo-classical style which was at this time becoming popular in Devon.[20] This is one of the earliest surviving Greek

Buckland House, Buckland Filleigh, North Front, 1810 (SS464094)

Doric revival houses and remains largely unaltered, probably incorporating the core of the earlier house. It is a Grade II* listed historic building.

This house, one of the largest in Devon, shaped in plan to the letter F, has its long side running east to west with the main entrance slightly offset facing north beneath a portico with four columns. There is another four-column portico facing east. In the grounds to the south is a lake, probably constructed during the ownership of William Fortescue before 1834. That Green undertook this work nearly 30 miles from Exeter showed his confidence in management and in delegating to skilled workmen. The construction of Hele Bridge, Hatherleigh, only three and a half miles away, appears to have been undertaken at about the same time.

2.72 ST DAVID'S CHURCH

When Green arrived in Exeter there was little inhabited property along St David's Hill. The existing church was described as 'small and irregular, consisting of a nave and one aisle without a chancel. The whole of the building is remarkably low as is the tower which is square containing 4 untuneable bells; the church is light, well seated and in good repair. Built in the fifteenth century.'[21] It was, perhaps, surprising that a new church should have been built at that time, especially by a Quaker.

A notice in the *Exeter Flying Post* tells us that the foundation stone of the new church was laid on Tuesday 4 June 1816. It was to be a Grecian building of the Doric Order with a spacious and lofty portico at the West End. 'It was designed and to be executed by James Green, Architect and Civil Engineer whose experience and abilities were well known to the public and left no doubt of it being completed in a manner to give general satisfaction.'[22] The west end was surmounted by an octagonal tower which later caused it to be dubbed the Pepper Box. Not unlike Wren's London churches, it particularly resembled Telford's 1793 St Mary Magdalen church at Bridgnorth, with which Green must have been familiar having worked with his father in that area.[23]

Green's tower was octagonal with eight Doric pillars enclosing the belfry and was surmounted by a rounded dome carrying a cross on the top. The tower stood on a square platform and at each corner stood a large Greek urn. A photograph of the interior can be seen in the *Transactions of the Devonshire Association*.[24] The present church was begun about June 1897 and during the demolition work on Green's structure a copper plate of considerable interest was found. It listed the committee responsible for construction and the architect of the 1816 church and so was placed on a wall near the entrance inside the new church.

2.73 FURSDON HOUSE, CADBURY

Between 1813 and 1818 George Sydenham Fursdon undertook a number of repairs and alterations to Fursdon House. In 1815 a library was added to the west wing and in 1818 the Ionic colonnade was added at the front of the house.[25] In the house museum is an account presented by Green dated April 1818.

St David's Church, Exeter 1816
    Jas Richards
    Dr by order of Mr Green, Architect
    Colonnade   £55 15s 7½d

The work on the library was carried out by Green and is a matching two bay extension with larger windows.[26]

2.74 KILLERTON HOUSE, BROADCLYST, NEAR CULLOMPTON

In January 1812 John Rennie wrote to Sir Thomas Acland, Bart., of Killerton House proposing a scheme of water supply from the river Culm at Culm John mill.[27] Some 860 yards of 6 inch diameter pipe would carry water up 200 feet to a reservoir on the hill which would be designed by James Green. The reservoir would be 160 feet long by 80 feet wide and 8 feet deep and would be filled in two days. Water could then be supplied to the house, offices, gardens and meadows.

The suggestion was not taken further and much later, in 1844, when the Bristol to Exeter railway was being built, a source of supply was found elsewhere at Halliwell or Holy Well.[28] This may well have been the first opportunity Green had of being noticed by Sir Thomas.

At this time the Exeter to Cullompton turnpike lay within a quarter of a mile of Killerton house and also close by was the Crabtree Ale House. This turnpike road rose to a height of 462 feet near Bradninch and included three gradients steeper than 1 in 7. Within a year a Bill was before Parliament and in May 1813 an Act was passed to provide a new road from Hazelstone north of Broadclyst and running north on the eastern side of the river Culm to Cullompton. The new road was made three-quarters of a mile east of Killerton and brought more privacy, something which became increasingly important as the road developed into the main trunk route from Exeter to Bristol. Sir Thomas did not become a trustee of this route at first and Green did not survey or build it but one cannot but wonder whether the general idea did not come from Rennie or Green.

2.8 FAMILY MATTERS

On Thursday 28 December 1815, the *Exeter Flying Post* recorded the death on Sunday 24 December 1815 of Thomas, aged 3 years, the infant son of Mr Green, civil engineer, St David's Hill.

Two years later, on 4 September 1817, the same newspaper carried an announcement that on Sunday 31 August 1817 the wife of James Green Esq., Civil Engineer, had borne a son. Neither the wife's nor the son's names were recorded in that brief announcement but in the list of civil engineers published in 1838 by the Institution of Civil Engineers we find an associate named Joseph Green of Exeter.[29] This was the son born in 1817. Joseph became a useful engineer and later held a post in Bristol Docks, afterwards working for James Rendel on the New River project in Wisbech.

As we shall see in chapter 5, James' son's Christian names were Joseph Dand. When Green signed his contract to build Thorverton Bridge for £4000 on 17 December 1811, his supporters were Joseph Dand of Sparkbrook, Birmingham in Warwickshire, Gentleman, and John Schofield of Birmingham, Gentleman. Obviously James Green was particularly indebted to the former.

2.9 TURNPIKE ROAD – EXETER TO PLYMOUTH

As the end of the decade approached, Green switched his attention from land reclamation to the rewards obtainable from the development of the turnpike roads and the construction of canals. As we have seen, he had established a relationship with the trustees of the Plymouth Eastern Turnpike before he became County Surveyor and now he approached the trustees of the three turnpikes responsible for the Exeter to Plymouth road with proposals for the realignment of the road to eliminate the steep inclines that made the passage of mail coaches, or any horse drawn vehicles, so difficult.

In a report to the trustees of the Plymouth Eastern division of turnpike roads, the trustees of the Ashburton division and the trustees of the Exeter turnpike Green completed a survey and produced plans and estimates for improvement 'the great object of which is, either to avoid altogether the hills by which travelling along this road is so peculiarly impeded, or so to lessen their ascents and declivities, as to give that degree of safety and expedition to travellers on this road, which is so essentially necessary on so important a line of communication, as that between the metropolis and one of the grand naval arsenals of the kingdom.' [30, 31]

Faced with a 1 in 7 incline east of Bittaford, 1 in 10 at Brent Harbertonford and 1 in 8 at Dean Clapper Hill, Green proposed realignment of the road so that the only steep gradient would be 1 in 15 at Whiteoxen Farm on the eastern side of the Harbourne stream. Here he proposed an embankment 18 feet high with the river in a stone culvert. To assist the movement of cattle another stone culvert was built as an underpass. Both of these long structures beneath the embankment survived until the main Exeter–Plymouth road was reconstructed in the 1960s and 1970s.

At this time Dart Bridge, Buckfastleigh, was only 9 feet between the parapet walls and Green pointed out that this width could be doubled by arches turned on each side, springing from the cutwaters of the piers. The bridge was widened in 1827.

At Bickington the descent to the bridge was 1 in 9 and the ascent towards the east was 1 in 7 so Green proposed a new line of route which reduced the gradients to 1 in 15. This improved road lies between the old route and the 1974 dual carriageway and is interesting for the bridge built by Green (and widened in later years) and for the high retaining wall on the approach from Ashburton. The retaining wall has had to be supported by sloping buttresses to resist the developing earth pressure on an otherwise relatively thin masonry wall.

Green considered the existing road from Chudleigh to Exeter through Shillingford as bad, being nothing less than a succession of very steep hills for nearly the whole distance. Gradients of 1 in 8, 1 in 10 and 1 in 8 rose to Haldon Hill while the descent to Exeter was 1 in 6 with an ascent of 1 in 8 at Clapton Hill. As the summit of Haldon Hill had to be attained in some way, Green proposed a new route which reached the summit a few chains north-west of the racecourse grandstand some 30 feet lower than the highest part of the old road. Gradients would be 1 in 18 on the Chudleigh side of the hill and 1 in 16 on the Exeter side where it descended towards the Exeter to Newton and Teignmouth road near its 5 mile stone. Although overall the distance from Chudleigh to Exeter was increased by a quarter of a mile, as the Exeter to Newton portion of the route was already maintained by the Exeter Turnpike trustees their overall

BRIDGES, LAND RECLAMATION AND ARCHITECTURE

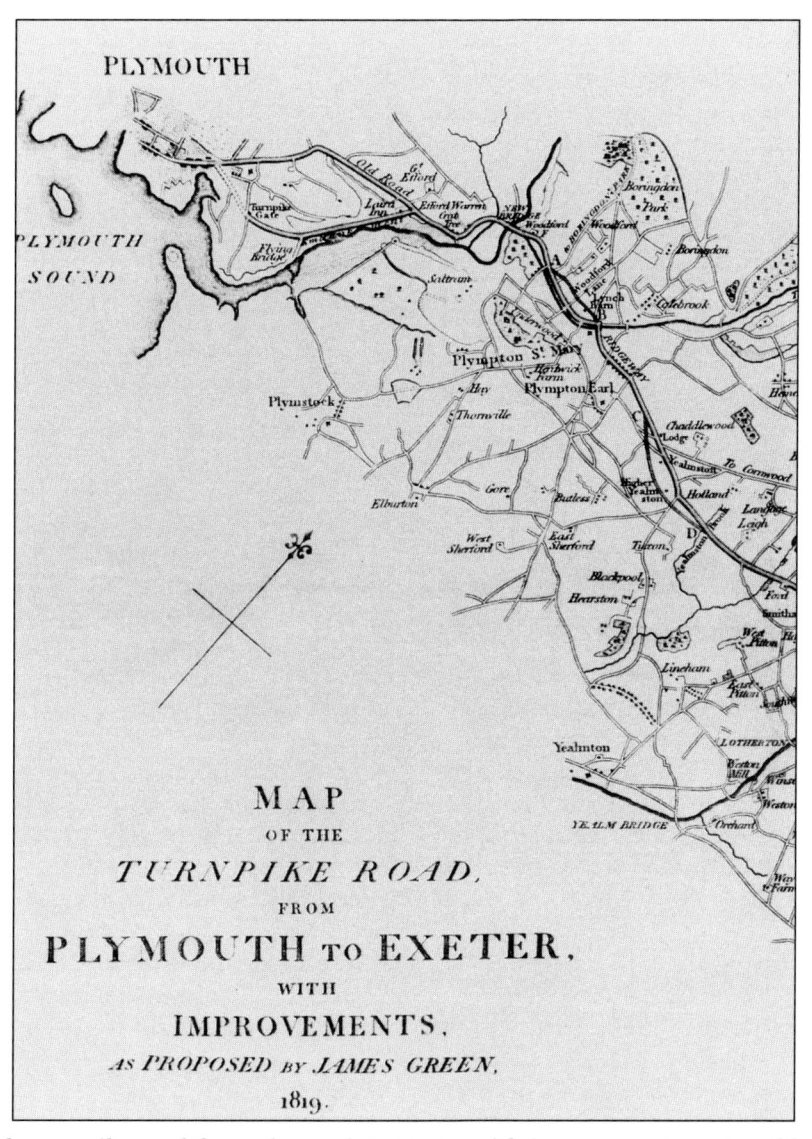

Extract from map of the turnpike road from Plymouth to Exeter with improvements proposed by James Green, 1819

highway commitment could be reduced by some 5 miles when the existing route no longer had to be maintained by them.

Eventually some 14 miles of road were realigned according to Green's proposals. His estimate for the work was £61,636. His report was dated 30 June 1819 and his plan was presented to the Clerk of the Peace on 15 January 1820, this time with the bypass to Kennford and the Buckfastleigh to South Brent length omitted.[32] Nevertheless work on these alignments was eventually carried out, the Buckfastleigh to South Brent length quite soon going by the evidence of the construction of the structures en route, and the Kennford bypass much later between the two world wars.

The Buckfastleigh to South Brent length was interesting from a highway engineer's point of view because of the grade separation of local traffic at Whiteoxen. First there was the underpass for cattle already mentioned. Second, at the top of the hill, the road was cut below an arch bridge which carried the local Rattery to Harbourneford road at ground level to save a descent and ascent for that traffic. This bridge was removed and reconstructed when the Exeter to Plymouth road was widened in 1966.

2.10 THE BUDE CANAL

Early ideas for a canal from Bude to Launceston surfaced in the 1770s when an Act was obtained and when John Smeaton surveyed a route. In 1793 Charles, Earl of Stanhope, commissioned two engineers to survey a line but he died at an early age. It was therefore the fourth earl who eventually supported the scheme for which James Green and the surveyor Thomas Shearm were commissioned in 1817 to survey a possible line for the canal.[33, 34]

Robert Fulton had already suggested to the third earl that inclined planes would be more suitable than locks for this scheme and Green took over this idea for the 362 feet rise from the sea to the river Tamar and also the idea of using small boats with wheels to run on the inclined planes. He turned these ideas into a proposal suitable for the demands of the district.

Unfortunately the scheme which was shown on Green and Shearm's plan of 1817 annoyed the Duke of Northumberland because it ran exceptionally close to his house at Werrington Park and so a revised plan of 1818 pulled the line away towards the river Tamar. An enemy had been made, and eventually the promoters had to settle for terminating the Launceston branch at Druxton.[35]

An Act of Parliament was obtained in 1819 and after a meeting of shareholders at Exeter on 19 July, work began on 23 July. James Green showed his commitment as the engineer by also becoming a shareholder with a contribution of £3000. How he obtained this sum of money is a matter of great interest for he had recently built his large house at Elmfield in Exeter while his initial salary as County Surveyor was just £300 per annum.

Green brought in navvies to excavate the canal from works upcountry as local labour had insufficient expertise. 35 miles of cut were required from

BRIDGES, LAND RECLAMATION AND ARCHITECTURE

Bude to the high ground near Red Post and thence south alongside the Tamar to Druxton, east to Thornbury near Holsworthy and north to the reservoir on the Tamar that would supply the water. Engineering works included six inclined planes, a sea lock at Bude, aqueducts over tributaries and an impressive earth dam at Alfardisworthy. For all this Green would direct the work as engineer, no doubt with the assistance of Shearm in setting out the works to line and level.

A Report to the subscribers of the proposed Bude Canal dated 14 April 1818 showed Green's breadth of vision.

> **The Tamar Lake Reservoir**
> I propose to make at the extremity of this branch on Langford Moor a Reservoir of some 60 acres averaging 12 feet deep. The whole body of the Tamar, beyond what is necessary to the supply of the mills may be turned into it, without any injury to any mill but that of Alfardisworthy, which may be built on another more advantageous spot. By this means it may be kept at times nearly full, except in very dry seasons, which rarely continue long.
> This Reservoir will contain upwards of 870,000 tons of water. I calculate that the method by which I propose to raise the boats will not require more than 2 tons of water for each ton of cargo.
> The Reservoir will therefore be equal to the supply of water, to pass a trade of at least 350,000 tons (making ample allowance for exhalation) which is much beyond whatever can be expected to be carried on this canal during the dry seasons of the year.

> N.B. The estimate for the whole of the above work totals £3714 0s 0d.

John Kingdon was appointed inspector or resident engineer and extracts from his diary show the state of the works as they progressed after the July 1819 commencement.[36]

22 October 1820
> Remained at Bude in the company of the chairman (G.C. Call) watching the effect of the high tide and seas on the breakwater and sea lock. Counted 340 men on the work.

6 December 1820
> Party of masons building a bridge near Hele Bridge where the course of the stream is finished. Parties at work on the Hobbacot incline and Marhamchurch incline planes and about 40 men excavating the sea lock channel.

14 December 1820
> The shaft at Marhamchurch incline for the waterwheel to work is fallen in. A gang of workmen is dealing with it.

27 December 1820
> One of the whims at Hobbacot at work and one shaft expected to be clear of water this night. Plane party formed and each shaft down about 70 feet and the adit driven about 60 feet.

2.11 TOWARDS THE NEXT DECADE

As 1820 approached, together with his thirty-ninth birthday, Green had established himself as a County Bridge Surveyor with many fine bridges to his credit. He had proved his expertise in the reclamation of land and had showed his sensitivity as the architect at Buckland House, Elmfield and St David's church. As a highway engineer he had initiated a series of schemes with his proposals for the Exeter to Plymouth road and now, with the Bude Canal, he was embarking on a career as a canal engineer where the quantity of civil engineering work would ensure further additions to his income – or so one would anticipate.

Unfortunately, Green's investment of £3000 in Bude Canal shares produced no return in his lifetime, the first dividend being paid in 1876 at 10s in £50 with only seven lesser dividends to follow. On the other side of the scales he would have received his fees as engineer for the work and thereafter he attracted commissions from other interested promoters because of his achievement. During the next decade he was busier than ever.

However, W. Buckingham, the fourth and final clerk to the trustees of the Exeter Turnpike Trust, writing in his history of the trust in 1883, said of Green that in his canal work he entered into speculative contracts and so died in reduced circumstances.[37] Certainly during the next 20 years Green was to carry out a large amount of civil engineering work in addition to looking after the county bridges, most of it successful, all of it showing the breadth of his expertise and his capacity to undertake extensive travel, but some of it demonstrating the hazards of being a civil engineering contractor.

CHAPTER 3
1821-1830
BRIDGES, CANALS AND TURNPIKES

3.1 INTRODUCTION 1821–1830

During this decade one important scheme followed another. With the construction of the Bude Canal in progress, Green was approached by the city of Exeter to advise them on the maintenance and improvement of the Exeter Canal, work that was to last throughout the decade. The county decided that too much money was being spent on bridge improvements and cut their expenditure considerably, but the turnpike trusts asked for more and more bridge construction to be planned and supervised by the County Surveyor so the list of new constructions was larger than for the previous 12 years. In 1820 Green proposed 6 miles of new road from Pocombe Bridge to Tedburn St Mary for the Exeter Turnpike Trust which promptly set this work in hand.

When in 1821 ideas were being revived for a canal from the Bristol channel to the English channel, Green proposed a tub boat canal from Creech St Michael to Beer. By 1824 opinion had shifted in favour of a full sized ship canal. Telford gave advice and a scheme was proposed from Stolford to Beer with Green signing the plans. There may have been a particular reason for this. At that time Thomas Telford was engineer adviser to the Exchequer Loans Commission and as it was Telford's duty to weigh the merit of any loan application there was scarcely an engineering undertaking of any moment in the country which did not at some time claim his attention.

L. T. C. Rolt has written that in considering the formidable list of projects to which loans were granted it is not always easy to determine the degree of Telford's participation. In many cases it would seem that he was content to approve a scheme on paper. In others he would send one of his assistants to examine and report upon the state of works before making his decision. Thus Alexander Easton was sent to Cornwall in 1823 to investigate the Bude Canal with its inclined planes. His confidential report to Telford disclosed a curious state of affairs not altogether to the credit of James Green, the company's engineer. Nevertheless the Bude Canal was granted a loan of £20,000.

Certainly the city of Exeter called Telford in on two occasions to comment on proposals by Green for the improvement of the Exeter Canal and on both occasions Telford supported Green wholeheartedly. It was in 1824 that Telford signed Green's application to join the newly formed Institution of Civil

Engineers and Green's obituary reports that Telford offered him the appointment of engineer for the government in the Ionian Islands (which, however, he declined).

In 1823–24 Green combined with Underwood, the Somerset County Surveyor, to produce plans for a new house of correction to stand alongside the county gaol in Exeter and Green was responsible for the construction of this £12,700 building. A proposal for a Liskeard to Looe Canal came to nothing in 1823 but in 1824 Green commenced the Torrington Canal for Lord Rolle together with the necessary mills. In 1823–24 he surveyed St Ives and Ilfracombe harbours with Joseph Whidbey and in 1823 he employed Richard Andrews to survey a route from the Flying Bridge at Laira to Ermington and Ugbrook at Lady Down. At much the same time the road into Totnes was straightened and Green provided the bridges for both these lengths.

Sir Thomas Acland asked Green to survey the roads around Stratton in Cornwall in 1824, the next year the Ilminster Trust asked him to report on their road to London and in 1827 he advised the Chard Turnpike Trust on a diversion that had been proposed by their surveyor that is now the road through Yarcombe. In that year he built a lake at Blachford Park, Cornwood, and made proposals for a harbour at Combe Martin. In 1830 he proposed, but did not carry out, a new road from Topsham to Exmouth.

Green does appear to have influenced the development of the road from Exeter to Barnstaple which the Exeter Turnpike Trust and the Barnstaple Turnpike Trust built down the valley of the river Taw. Green designed five bridges which were built in 1830 and for which he was paid by each trust.

The Exeter Canal reached its new outlet at Turf in 1827 and at the end of the decade the canal basin was also finished to great excitement in the city. Suddenly the Devon magistrates realised that the large salary that they had awarded Green in the previous decade, together with his claims for 'extras' for work on the house of correction had amounted to a large sum, and they wondered whether he was being overpaid!

3.2 THE COUNTY BRIDGES AND THE COUNTY SURVEYOR

3.21 DEVON QUARTER SESSIONS ADMINISTRATION

The court of Quarter Sessions was ever anxious to control bridge expenditure and at Easter 1820 it required the treasurer to record separately the expenditure on each county bridge. The following Easter Green's appointment was continued for one year at £550 per annum and the court received a report from the committee of expenditure.

The report for the seven years 1812–1818 inclusive recorded the committee's comparison of Devon with Somerset and Cornwall. In Devon for 259 bridges the county spent £226 per bridge, a total of £57,412. In Cornwall for 71 bridges the county spent £53 per bridge, a total of £3,709. In Somerset for 96 bridges the county spent £77 per bridge, a total of £7,447. All include the costs of roads belonging to them approaching the bridges.

The committee of expenditure recommended as follows:
1. The Act of Parliament 31 Geo II entitled 'An Act for the more early and speedy repair of the Public Bridges within the county of Devon' should be repealed, and Devon should conform with the general law.
2. A bridge committee should be appointed comprising two or more magistrates from each division of the county.
3. An abstract of accounts should be annually audited by the bridge committee and published in the provincial newspapers.
4. The magistrates of each division of the county should form a divisional committee for the management of the bridges in the division.
5. For the special care and superintendence of each bridge, two or more magistrates residing nearby should be annually appointed by the court pursuant to 52 Geo III c110.
6. The Clerk of the Peace should advertise for tenders for repairing the superstructures of the public bridges.

The cost in pounds of the work in Devon over the years 1812–1818 was detailed. (see table below)

	1812	1813	1814	1815	1816	1817	1818	TOTAL
New bridges	3605	8510	5483	6985	2490	1260	900	29,234
Purchase and damage done to land	105	800	500	132	729	170	23	2,462
Repairing bridges	3637	2692	2808	2256	1849	2349	3162	18,752
Repairing roads ove bridges	133	221	236	505	838	839	756	3,531
Surveyor's salary	400	400	400	552	577	550	550	3,430
TOTAL	7881	12624	9428	10431	6483	5170	5391	57,412

The figures showed a marked reduction in rebuilding costs through the seven years after schemes like Cowley Bridge had been completed, but were accompanied by a noticeable increase in the cost of repairing the approach roads to the bridges as that system became more effective.

Green's explanations did not convince the Quarter Sessions and expenditure began to change thereafter. Green was unable to convince his magistrates that Devon's rivers were longer, with higher catchment areas than the other two counties, requiring much more capacious bridges to pass flood waters. Perhaps his other 'fault' was that his bridges were sound and adequate for all the floods to come in the nineteenth and twentieth centuries.

At Epiphany Sessions 1822, Green's appointment was renewed for a further year and at Easter a committee of magistrates was set up to implement the recommendation of the committee of expenditure. The bridge committee produced a list of bridges in the divisions with the relevant magistrates, but the list was incomplete, suggesting that not all magistrates were enthusiastic about their bridging duties.

At Midsummer, Palmer's Bridge in Cullompton was repaired as a county bridge, while at Michaelmas it was ordered that William McAdam should be paid in advance when required for county bridge roads within the limits of the Exeter Turnpike Trust, provided the payments were kept within the yearly limit.

At Easter 1823 the bridge committee had a resolution adopted by the court that the County Surveyor should, in future, certify a distinct and separate bill for repairs or rebuilding work done for each particular bridge in order that no charges to two or more bridges be included in one bill. Green's appointment was continued for a further year, 'he having declined to make an engagement for a shorter period.' At Midsummer it was ruled that persons contracting with the county for bridgeworks should not contract with persons under them for erecting the work while at Michaelmas the County Surveyor was given notice of a review of his appointment at the next Epiphany Sessions.

Obviously there were differences between Green and the magistrates, possibly because of his intense involvement in other works on canals in the county. A report stated that 'the salary attached to this officer is in the opinion of the Committee (of Expenditure) a very liberal one, but they concur with the bridge committee in thinking it is not an excessive remuneration for duties now appertaining to it, when performed by a person of such eminent professional talents as the present surveyor; and merely for the purpose of some saving in salary they do not venture to recommend the appointment of a person of inferior abilities.'

At Midsummer 1824 Green was reappointed from the previous Easter at £550 per annum to superintend buildings as well as bridges but it was significant that he must give notice to the justices appointed before making a survey of any bridge so that the justices could attend the inspection.

At Michaelmas it was ordered that the lists of bridges and magistrates settled by the bridge committee should be printed and transmitted to every justice in the county and that every division should select particular magistrates to superintend each bridge and send this information to the Clerk of the Peace. The work on roads approaching bridges had been contracted out by advertisement in April 1821. Now it was decided to include the parapets or guard rails and railings of the bridges and the following advertisement appeared.[1]

> Devon County Bridges
> Any person or persons willing to undertake the repairing and keeping in repair for 5 years from the next Michaelmas Sessions, the parapet or guard rails and railings of the several bridges maintained and repaired at the cost of the county of Devon, together with the county Roads or roads over and adjoining the said bridges, may see the list and several divisions of the said bridges and the Specification of the several works to be performed on applying at the office of the Clerk of the Peace for the county of Devon in the Castle of Exeter and they are desired to deliver sealed tenders for performing the works at my office on or before 14 October next.
> 2 September 1825
>
> R. Eales
> Clerk of the Peace

Green was devolving a portion of his workload that was both time consuming and required frequent attention, leaving himself time to consider larger issues. This process would be taken a stage further in 1831 when contractors would be instructed to keep a general watch over the substructures as well. At the Michaelmas Sessions on 18 October tenders were referred to the bridge committee who were requested to enter into the contracts for the repairs. Some contractors would be responsible for a group of bridges and their annual payments would be made at each Michaelmas Sessions.

At Easter 1826, two contractors, Curson and Pearce were reported for neglecting their contract and after being given time to respond their contract was rescinded at Midsummer.

By Midsummer 1828 a new committee had been appointed to review and amend the list of county bridges and the magistrates appointed to superintend them. It also had to consider highway legislation then proceeding in Parliament. By Epiphany 1829 the bridge committee had reported new arrangements and the lists and letters to magistrates were printed, but at Easter when the lists were recorded in the minutes, only 173 bridges appeared, so the organisation was creaking. At the next Sessions the clerk was instructed to write to the defaulting divisions and at Epiphany 1830 the committee was reappointed to examine the divisions. By Michaelmas the contracts had expired and the bridge committee was instructed to make new arrangements.

Meanwhile, at Easter 1830, notice was given to the County Surveyor that the county would determine their engagement with him at the expiration of one year and the salaries of the officers of the county would be considered at Michaelmas 1830. Trouble was afoot.

Clyst Honiton Bridge 1821 (widened 1969) (SX985935)

3.22 BRIDGES RENEWED AND REPAIRED 1821–1830

Whereas Emmett's Bridge in 1811 had been an exception, in this decade the County Surveyor's supervision of the design and construction of bridges extended rapidly to bridges promoted by the trustees of the turnpike trusts, or other individuals or groups of people. This had been the case for Brightley Bridge, Okehampton, mentioned previously. It was necessary if a bridge were to be adopted as a county bridge by the court of Quarter Sessions for the provisions of the 1803 Act 43 Geo III cap 59 to be observed; namely that the County Surveyor should approve of the plan and specification of the bridge and, being satisfied that it had been properly constructed, should report this to the court.

Although responsibility for bridge works always rested with the local magistrates this 1803 Act gave the County Surveyor the opportunity to ask the proposers if they wished him to prepare the plan and specification and, if they agreed, to charge them for this service. One might say that the proposers could hardly disagree as there was unlikely to be anyone else in the area with the professional knowledge anyway! Where his fee has been recorded, Green seems to have charged £30 per bridge.

On 10 August 1821 the Countess Wear committee of the Exeter Turnpike Trust were considering a new drawbridge over the canal, the widening or rebuilding of the adjacent stone bridge over the Alphington brook and improving the approach 'avenues'. They ordered this work to be carried out by James Green to an original estimate of £800.

For the county, New Bridge near South Brent over the river Avon (now Avonwick Bridge) was built for a cost not to exceed £600 in 1821. The three-span Clyst Honiton Bridge was also built for £1200 and this important bridge carrying the Exeter to Honiton road became part of the trunk road A30 and was carefully widened by the Department of Transport in the early 1970s with two footways. Clifford Bridge over the river Teign was widened in 1821 by springing arches from the cutwaters, it having formerly been little wider than a packhorse bridge. It is now scheduled as an ancient monu-

Clifford Bridge, widened 1821   (SX780987)

Sidmouth Bridge 1821   (SY128878)

ment. Sidmouth Bridge was advertised for contract by James Clarke Esq. of Sid Abbey, Salcombe, and appeared in the 1831 bridges list, confirming Green's involvement.

At Epiphany Sessions 1822, James Pitman Esq., a J.P. for the county, in pursuance of 43 Geo III c59 requested the County Surveyor to supervise construction of a bridge in the parish of Dunchideock in the road leading from Exeter through Dunchideock to Chudleigh. The order was made without a cost so the work was financed by Pitman and others. It is still there, a small span in a deep gulley.

The only bridge ordered by the court in 1823 was Westwood, on the Crediton to Bickleigh road, but this was a small structure and was removed during road improvements in the 1960s. About this time the Bude Canal was nearing completion and a bridge to carry the coastal road was built at Hele, below Marhamchurch. It can now be found just alongside a new trunk road bridge carrying the A39.

In 1824 the new Exeter to Tedburn St Mary road was being built by the Exeter Turnpike Trust on an

Eggesford Bridge 1824   (SS683114)

alignment proposed by Green. While there are no large structures on this length, a pointer to the future was the underpass for cattle moving to and from Great Huish Farm. The magnificent five-arch Beam aqueduct was also being built by Green for Lord Rolle's Torrington Canal and this has since been converted as a road bridge to give access to Beam House.

At this time Green was building Eggesford Bridge over the river Taw, a combined bridge and weir, for the magistrate, the Hon. Newton Fellowes,[2] later to become Lord Portsmouth, who, within a few years, was active in persuading the Exeter and the Barnstaple turnpike trusts to build a new road down the Taw valley. (see 3.48)

The Plymouth Eastern Turnpike required new bridges for its route from Laira to Lady Down (see 3.44) and the court agreed to share the cost with

Great Huish Farm underpass 1824   (SX828937)

the trust for Yealm Bridge and for Erme Bridge. This latter was really two bridges which are now called Erme and Little Erme, with the river Erme flowing under the first and the Erme's tributary, the Ludbrook, under the second. Also, the Reverend Mr Harwood and others wished to erect at their own expense a bridge over the river Culm and asked the court to direct the County Surveyor to supervise its construction.

Another request in 1824 came from Sir Alexander Hood for the County Surveyor to supervise the

Long Bridge, Cullompton 1825 *(ST025077)*

erection of a bridge in the parish of Thorncombe over the river Axe. This bridge, known as Winsham Bridge, was shared between Somerset and Devon and its construction was reported by Green as approved in 1828. Boundary changes have since placed Thorncombe in Dorset, so Devon is no longer involved.

Small bridges at Hynah on the Hennock to Christow road and at Fulford Water on the Tiverton to Cullompton road and a larger bridge over the river at Dawlish were promoted by the county and finally at Michaelmas 1824 the reconstruction of Gosford Bridge over the river Otter was approved at a cost not to exceed £1200.

The year 1825 began with Chercombe Bridge near Newton Abbot being ordered for £250 and a substantial rebuild of Bovey Bridge for the same sum. Slapton Bridge was ordered as a wooden structure for £150 but as it was almost immediately washed down in a flood it was ordered to be rebuilt in stone. For the Ashburton turnpike trustees Green was ordered to examine the plans for a new bridge over the river Lemon on the line of the new road he had proposed in 1819 at Bickington. This procedure was repeated for a new bridge over the river Avon at Loddiswell and another over the river Harbourne, 300 yards north of Sandwell. There was also an un-named bridge for the Torquay Trust and at this time the direct road from Newton Abbot to Torquay was being built with a bridge over the Aller brook and a tunnel at Kingskerswell.

In 1826 Long Bridge at Newton St Cyres was taken down and replaced by a culvert, Clyst Hydon was

Gosford Bridge 1824–25 (SY101970)

Clyst Hydon Bridge 1826 (ST036015)

rebuilt at a cost of £300 and Tinhay Bridge approved by the county for £1200, it being agreed that the advertisement should be put in the *Swansea Journal*. For the Bideford Turnpike Trust, Green built his first cast-iron span at Pillmouth, near Landcross, for which a drawing of the 50 feet span remains. This bridge has since been rebuilt in 1926 and 1993. Green's drawing of August 1826 shows the cast-iron rib 2 feet 3 inches deep at the centre and 6 feet 9 inches deep at the ends resting on a quadrant of masonry to give both support and resistance to the thrust. Lateral connection with other ribs was provided by two rods, 18 feet from each end.

At Epiphany Quarter Sessions 1827, Green produced proposals for rebuilding Newton Poppleford Bridge and Otterton Bridge, both over the river Otter's lower reaches, the first priced at £2500 and the second at £1700. There seems to have been continual disagreement among the magistrates on whether to proceed at Newton Poppleford and on this occasion it was agreed to let that proposal lie on the table. Otterton Bridge, the most downstream of Green's six three-span bridges over the river Otter, was built with smaller spans than the others and even in recent years the approach roads have flooded in heavy run-off. However the court approved this proposal and work proceeded.

At Otterton a balance was struck by building a smaller bridge and leaving the approach road at the flood plain level. At Newton Poppleford the solution for a more important road, namely of building a bridge large enough to take all forseeable floods and raising the approach roads above flood plain level, was not to be achieved until 1841, and even this was not good enough for the catastrophic

Tinhay Bridge 1827   *(SX392851)*

Otterton Bridge 1827   *(SY079838)*

James Green – Canal Builder and County Surveyor

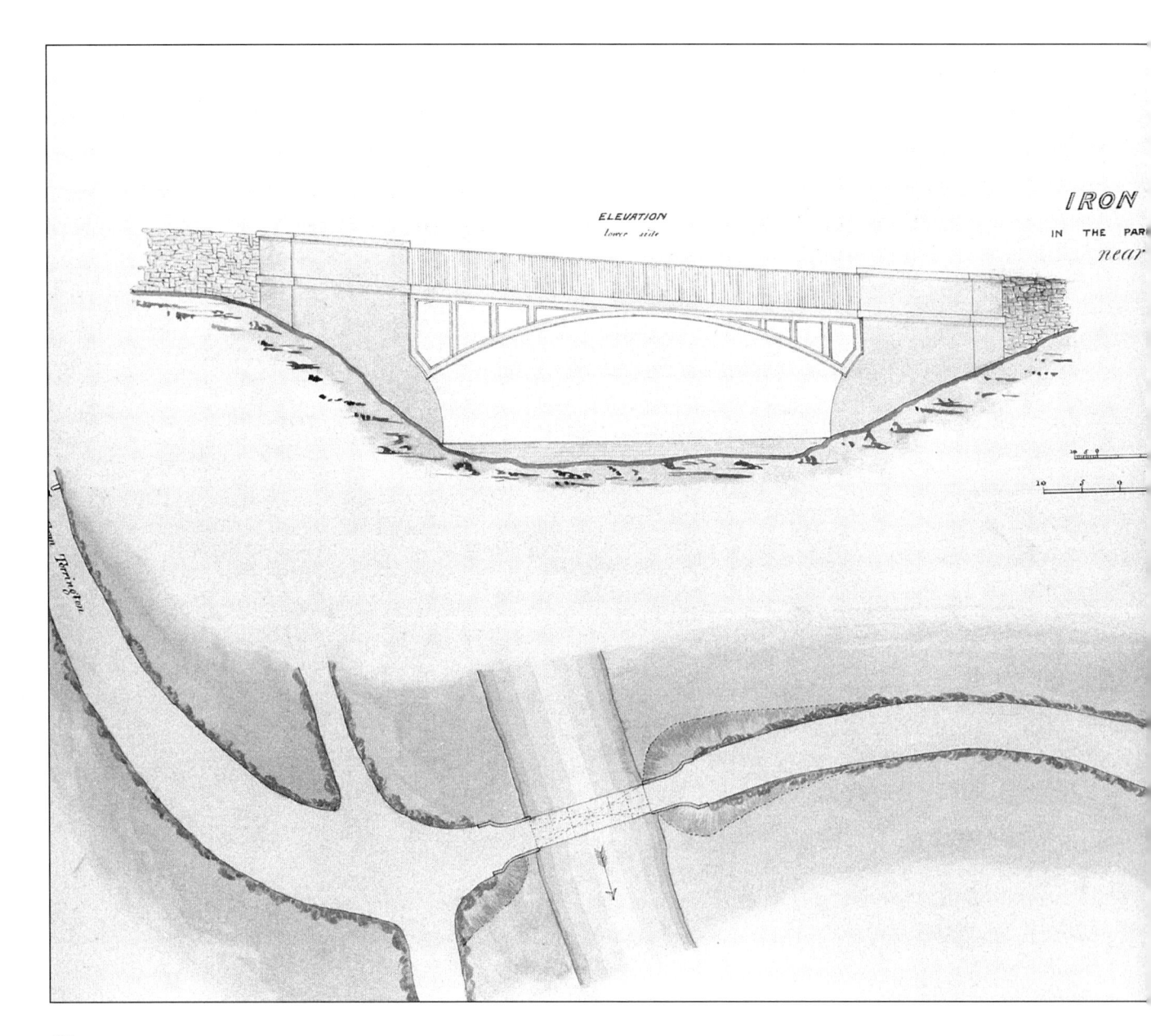

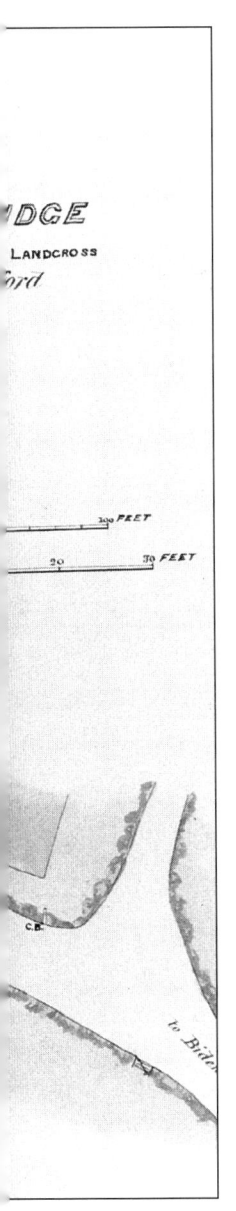

Iron Bridge, near Landcross, plan and elevation 1826 (SS695014)

flood of 1968, the worst for 600 years, which flooded the approach road across the flood plain.

At the next Sessions the court agreed that the earliest opportunity should be taken to rebuild Newton Poppleford Bridge. After a delay to consider alternative sites it was also agreed to proceed with Tinhay Bridge on the site of the existing structure. Leave was given to the trustees of the Totnes Trust to alter and improve Harbertonford Bridge subject to the approval of the County Surveyor. Because the mill leat at Newland Bridge was within the county bounds the court provided £110 for a new arch over the leat. Discussion during the latter part of the year concerning the site for a new bridge at Sowton Barton was resolved when it was decided to provide £700, and the following advertisement appeared in the *Exeter Flying Post* on 10 April 1828:[3]

Devon County Bridges

Any person or persons desirous to contract to build a new bridge over the river Teign near Sawlon Mill between the banks of Dunsford and Bridford may see the plans and specification of the works to be preferred at the house of the Rev. George Gregory at Dunsford and may deliver sealed tenders for the performance of the work at the office of the Clerk of the Peace at the Castle, Exeter, on or before Thursday 17 April next. The tenders will be taken into consideration and decided on at the Castle, Exeter on Friday 18 April. Security will be required for the due performance of the contract.

Only seven days in which to deliver a tender! The Easter Sessions were already in progress, but the award of the contract is not recorded there.

However, at Easter 1828 the county made a contract with John Spearman to rebuild Dipper Mill Bridge over the river Torridge for £799 and approved the provision of £300 for the rebuilding of Kennford Bridge. At Michaelmas the County Surveyor reported that Wooley Bridge near Beaford had been built in a substantial and commodious manner and so this bridge was taken over by the county.

New Bridge, Kingsteignton, was in trouble at the end of 1828. Green had built this bridge in the early rush of construction in 1810 and while the waterway capacity was quite reasonable for its situation with three spans of 20, 22 and 12 feet, it was only 10 feet 6 inches clear between the parapets. At Michaelmas it was declared by some magistrates to be in a ruinous state and the County Surveyor was ordered to inspect it and report, and at the next Sessions he was instructed to prepare a new plan and estimate. This was produced at Easter 1829 and the estimate, no doubt for a full width bridge,

Dipper Mill Bridge 1828 *(SS438065)*

was £2500. A motion to rebuild was refused and the bridge remains to this day.

Also in 1829 at Epiphany Sessions several parties proposed to erect a new bridge at Bampton over the river Batherm, a tributary of the Exe. The County Surveyor was ordered to inspect the plans and specifications and to report when all was complete. At Easter there was notice from both Exeter and Barnstaple Turnpike Trusts that each wished to build two bridges on the new line of road from Exeter to Barnstaple. Here we know that the County Surveyor, ordered to inspect plans and supervise construction, actually received £60 from each of the trusts for his services. For Barnstaple he had second thoughts on both Little Dart and Newnham Bridges and produced bridges with larger waterways than shown on his plans. For the Exeter Trust, besides the bridges at Lapford and Bury, there was also a mill leat bridge at Bury.

That Easter it was also ordered that Harbertonford Bridge be rebuilt. Materials for the bridge would be given to the Totnes Turnpike Trust on condition that they built a new and substantial bridge to the County Surveyor's specification.

At Easter 1830 the County Solicitor reported the liability of the county to repair Topsham Bridge and this report was accepted by the court. The county had had a Topsham Bridge on its list in 1809, the bridge over the river Avon on a route from Plymouth to Dartmouth. It appears, therefore, that the county was now taking over the bridge over the Clyst at Topsham on the road maintained by the Exeter Turnpike Trust.

Bampton Bridge, 1829 (SS959221)

At Midsummer the magistrates agreed to spend £150 on Wilmington Bridge if the Axminster Turnpike Trust would spend £50 on the approaches, while in the north it was agreed that £120 would be allowed to the Barnstaple Turnpike Trust if the County Surveyor's plans for rebuilding Winters Bridge were used. For Topsham Bridge £50 was now allocated for repairs, and the County Surveyor was again ordered to view Newton Poppleford Bridge. The need to rebuild Polson Bridge over the river Tamar on the county boundary near Launceston was becoming urgent and the justices and the County Surveyor were instructed to meet Cornwall's representatives. At Michaelmas a report from the magistrates was accepted and the County Surveyor was instructed to prepare plans in conjunction with the surveyor of Cornwall.

So ended a decade of building bridges for the county and the various turnpike trusts, a busy time

Little Dart Bridge 1830 (SS668137)

Lapford Bridge 1830 (SS727080)

Newnham Bridge 1830 (SS660174)

Bury Bridge 1830 (SS738068)

improving the more important roads to the standard needed to meet the increasing use of heavier and wider wheeled vehicles.

3.23 THE COUNTY BUILDINGS

At Easter Sessions 1821 there were a number of building matters in Exeter to concern the court. Green reported that the sheriff's ward had been built according to the contract and that since the completion of the contract several additions and improvements had been made resulting in a balance due to Messrs Cornish and Son of £562 4s 10d. The treasurer was ordered to pay.

The expenditure committee reported on expenditure on the buildings between 1812 and 1818 as follows:

Castle	£2,919	3s	6d
Sheriff's Ward	£1,483	15s	4½d
Gaol	£3,033	4s	9d
Bridewell	£6,240	9s	9½d
Total	£13,676	13s	5d

This was less than one quarter of the expenditure on bridges already noted, but to it had to be added Cornish's work on the new sheriff's ward since 1818. The Committee made the following recommendations:
1. All repairs should be done under contract. The County Surveyor to report at each Sessions on the state of each building.
2. When any alteration or improvement was required the County Surveyor should prepare an estimate for the court. Work would not be carried out until the court made an order at the next Sessions, excepting improvements that would not admit of delay.

Consideration was also given to installing machinery at the Bridewell similar to that employed at Bury St Edmunds. A committee was instructed to make enquiries and they were advised by Mr Cubitt, the engineer, that a sum of £2500 would be required for machinery for employing 84 men and for erecting a mill house and working sheds. This treading wheel was completed by Michaelmas 1822.

The County Surveyor was ordered to erect a necessary wall to enclose a piece of ground 30 feet square behind the north wall of the gaol required as a burying ground for malefactors executed at the New Drop. Green was also to make alterations to the Nisi Prius Court to deaden the echo. In 1823 he produced a plan for a more convenient withdrawing room and exit from the judges' chairs at the court which was adopted and £240 allowed when it was advertised for tender.

The previous Midsummer a committee had been appointed to consider an Act of Parliament passed relating to gaols and houses of correction and the County Surveyor had been asked to present a report on how the prison buildings could be rendered adequate to conform with the Act. This he did and reported at Michaelmas 1823 with two alternative schemes, one to cost £23,412 and the other £19,065. Sir Thomas Dyke Acland, the chairman, commended Green on his clear and able report, combining the gaol and the house of correction into one united prison under one keeper. In refining these designs, a reduction of some £6000 was made by accepting a recommendation of Sir I.C. Hippisley, Bt., a Somerset magistrate, who had accomplished a great enlargement of the prison at Shepton Mallet. This had been achieved at comparatively small cost by availing himself of the labour of the prisoners and he recommended the services of Mr G.A. Underwood, who was the Somerset County Surveyor.

Green and Underwood produced in January 1824 a plan and estimate of £12,700, with the recommendation that the employment of prisoners would produce further savings. This would enable 255 male and female prisoners to be confined to the house of correction and 151 male and female prisoners to be confined to the Gaol.[4]

An abstract from the estimate follows.

	£
Raising the House of Correction Buildings	2389
Making the proposed alterations in the Gaol and including the female prisoners from the House of Correction	6408
Additional buildings for laundry, washhouse, etc.	616
Infirmary, baths, etc.	3300
Total	12713

At Easter 1824 the court decided to pay Underwood £87 16s 0d for his work on this report and to pay Green £40. Green did not appreciate the difference although his salary already included an element for work on the county buildings and by Michaelmas he had persuaded his employers to vote him the balance of £47 16s 0d in addition. At this Sessions it was ordered that the plans should be carried out and that Green should prepare a specification so that the Clerk of the Peace could advertise for tenders.

Richard Eales, the Clerk of the Peace, turned Green's comparatively straight-forward specification into a 30 page document, *Articles of Agreement*, of great detail and complexity, together with a contract consisting of four large sheets of paper.[5] The following Easter, 1825, it was agreed that a contract should be made with Mr Robert Cornish, junior, for the alterations to the gaol and the house of correction in the sum of £13,489 17s 0d. In addition the committee should take into consideration the services of the County Surveyor in preparing the working plans and specifications.

Again, Green seems to have been pushing for extra remuneration for what was no more than his duty, but by Michaelmas 1825 a decision had been made to pay him £120 for his trouble and expenses. Furthermore he would be allowed 200 guineas per annum to supervise construction, from which fund he was to pay a skilful and competent clerk of works, with Green to be considered solely responsible to the magistrates for the due and proper performance of the work.

This allowance would begin from the date of employment of the clerk of works. Green would also be paid a sum not exceeding £10 to defray the expense incurred by him in obtaining the services of that officer previous to his starting his duties in Exeter. Meanwhile at Michaelmas 1824 a contract

House of Correction, Exeter

had been made with Mr Robert Greenslade of St Thomas to keep in repair the castle for £69 per annum and the sheriff's ward for £70 per annum.

When Cornish began his contract for the prison is unclear, but assuming that it was sometime after Green received permission to employ a clerk of works it would have been before Christmas 1825.

By Midsummer 1826 alterations to the gaol plan had been proposed by Mr Cole, the keeper, and referred to the committee, the visiting justices saying that there was too much space for female, and not enough for male, prisoners. Then, at Easter 1827, the gaol committee reported that it was unnecessary to make any addition to the house of correction and so that part of the plan was abandoned. For his part Green recommended a boundary wall improvement, sinking a new water tank and making a coal cellar.

By Epiphany 1828 alterations at the gaol were nearing completion and notice was given to the clerk of works that he would no longer be required at Easter, as the new buildings were complete and only the boundary wall remained to be finished. At Easter it was noted that of Mr Cornish's contract for £13,489, £13,900 had been paid already. New work not included in the contract needed to be measured and Mr Ayers of London was appointed to do this. At Michaelmas it was recorded that the arbitrator now found the total work valued at £17,686 12s 4d and as £14,400 had already been paid the balance due was £3286 12s 4d. Mr Ayers, the arbitrator, received a fee of £250.

In 1829 Green was paid £42 for two plans of the complete prison, one to go to the Secretary of State and the other to be kept in the Grand Jury Room at the castle. Green was also paid £100 on account of alterations made to plans found unsatisfactory and abandoned.

In 1830 at Easter Mr Fulford junior gave notice that he considered that all supplies should be contracted for at fixed periods and repairs to county buildings should be performed by contractors.

At Michaelmas the County Surveyor was instructed to prepare, in conjunction with the County Treasurer, a specification for the repair of the county buildings which would be suitable for a period of 14 years.

At Easter Quarter Sessions 1830 a letter from Colonel Taylor was read saying that the Newton Abbot Bridewell had been sold to the county. Colonel Drake said that the property was in a ruinous condition. Mr Green was ordered to look at the premises and Mr James (treasurer) to report at the Midsummer Sessions as to the best way for the county to dispose of it.

At the Midsummer Sessions Green reported that the dwelling house held by Mr Hearder, though old, could not be considered to be in a dilapidated condition and that there could be no objection at the expiration of the term to receive it at the hands of the lessee.[48] At Epiphany Sessions 1831 it was ordered that the county property at Newton Abbot be left at a rack rent agreeable to the county solicitor.

3.3 CANALS

Green's involvement with canals took off with the construction of the Bude Canal and thereafter he pursued the development of the tub boat canal for the west country. He believed that the smaller boats of 4 to 8 tons capacity with their wheels were best suited for the inclined planes that were essential for coping with the great changes of height above sea level on the routes from the sea inland.

Bude and Torrington Canals were quickly completed whereas schemes for Liskeard and Looe, Newton Abbot, the Western Ship Canal and the approach to Cardiff came to nothing. The Exeter Canal improvements were his greatest achievement lasting to the present day and capable of carrying ships of over 400 tons – very different from his tub boats!

Working with Telford on the Western Ship Canal proposals and having his Exeter Canal work checked by him, increased Green's reputation, while giving evidence to a House of Commons committee in 1825 extended his experience. This was faithfully reported in the Exeter press.

3.31 EXETER CANAL

Green's work on the Exeter Canal continued throughout the decade. In October 1820 when he first reported, the canal was still as it had been since the 1698 improvement which had lengthened it to the tideway opposite Retreat House. This lower sluice was approximately below the point where the motorway M5 crosses the river today. The distance of lower, or Trenchard's, sluice from double lock was nearly two miles and the lower sluice was only a pair of single gates so the loss of water in passing vessels at anything but very high water was immense.

The 1820 recommendations were as follows:[6]
a. Trenchard's sluice should be rebuilt as a regular lock with the cills lowered to river bed level;
b. the canal should have a uniform depth of 10 feet;
c. the cills of the gates of double lock should be lowered;
d. culverts should be constructed under the canal to drain the land cut off from the river; which land was fed with water from the Alphington brook.

The work proceeded and in March 1824 Green prepared a further report, this time proposing that the canal should be extended to Turf, about 2 miles lower down the river where vessels drawing 12 feet of water could navigate at all tides.

Thomas Telford was engaged by the city to report on Green's proposal, which he did in July 1824 confirming the route to Turf and also the building of a regular lock in place of King's Arms sluice where the canal re-entered the river above Trew's weir.

Green next reported in March 1826 suggesting that at least the lower part of the canal should be deepened to 12 feet and that banks through parts of the canal exposed to the tides should be raised 3 feet to guard against recurrences of the extraordinary tides of November 1824. If the banks were raised to that point a depth of 15 feet of water would be obtained from Turf to double lock.

During the previous alterations to double lock the cill and gates had been replaced to allow a depth of 15 feet. It only remained to deepen the canal between double lock and King's Arms sluice. Unfortunately it would be difficult to lower the river bed to the required depth between King's Arms sluice and the quays because of the solid sandstone rock below river. It was therefore proposed to form a basin, independent of the river, and connected to the canal below the King's Arms sluice.

Once again the report was submitted to Thomas Telford who pointed out that deepening the river

James Green – Canal Builder and County Surveyor

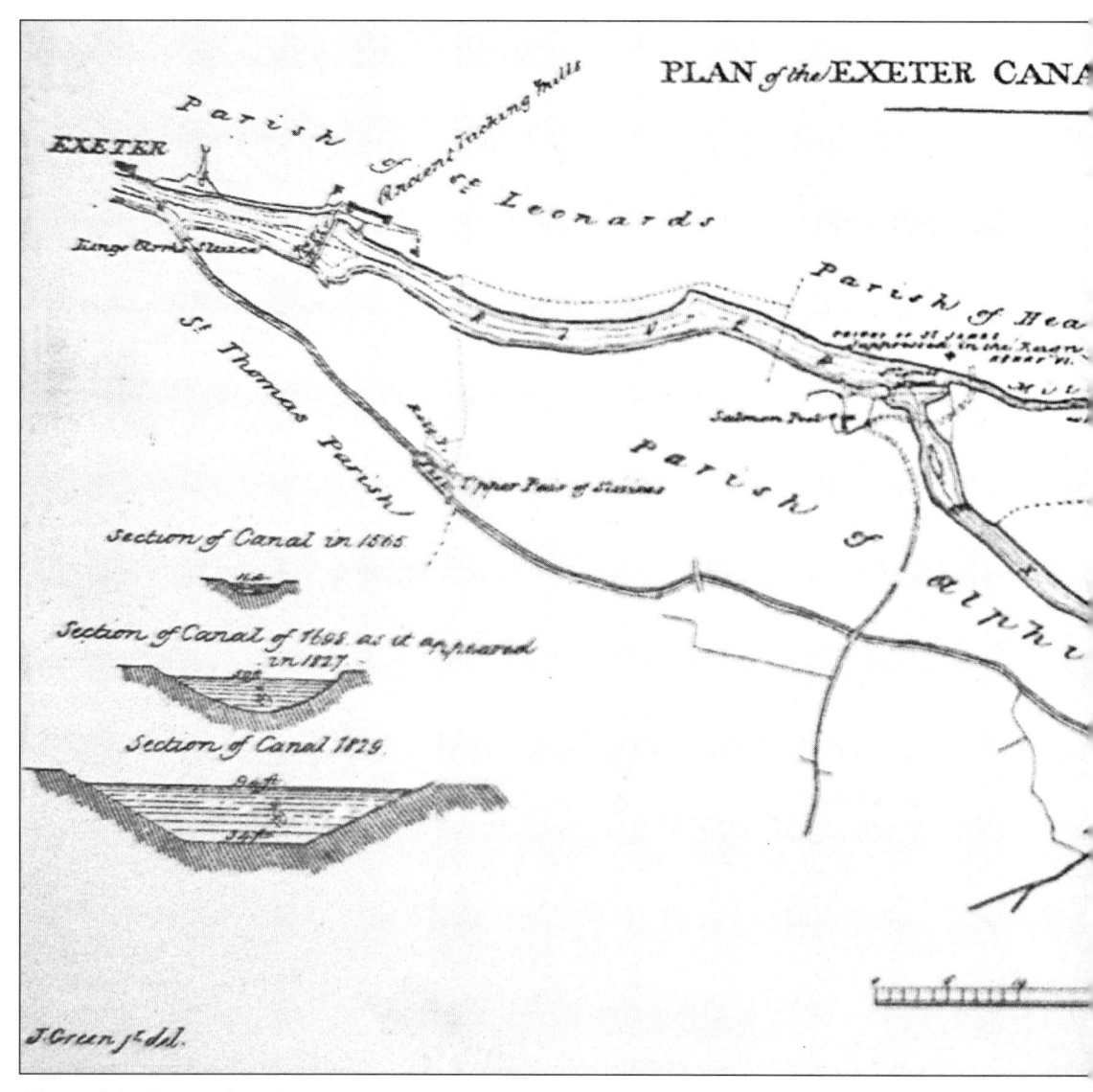

Plan of the Exeter Canal

BRIDGES, CANALS AND TURNPIKES

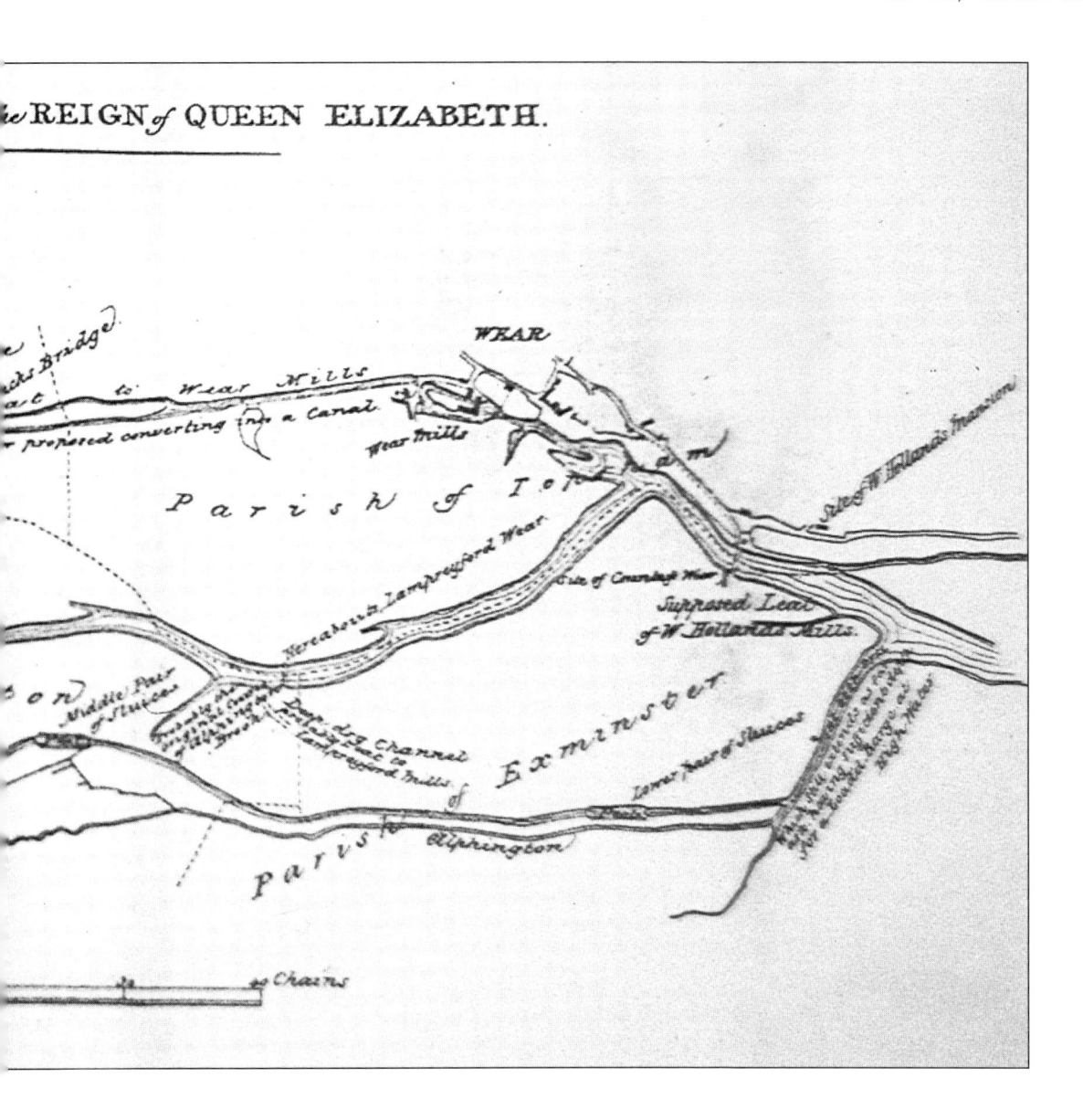

the REIGN *of* QUEEN ELIZABETH.

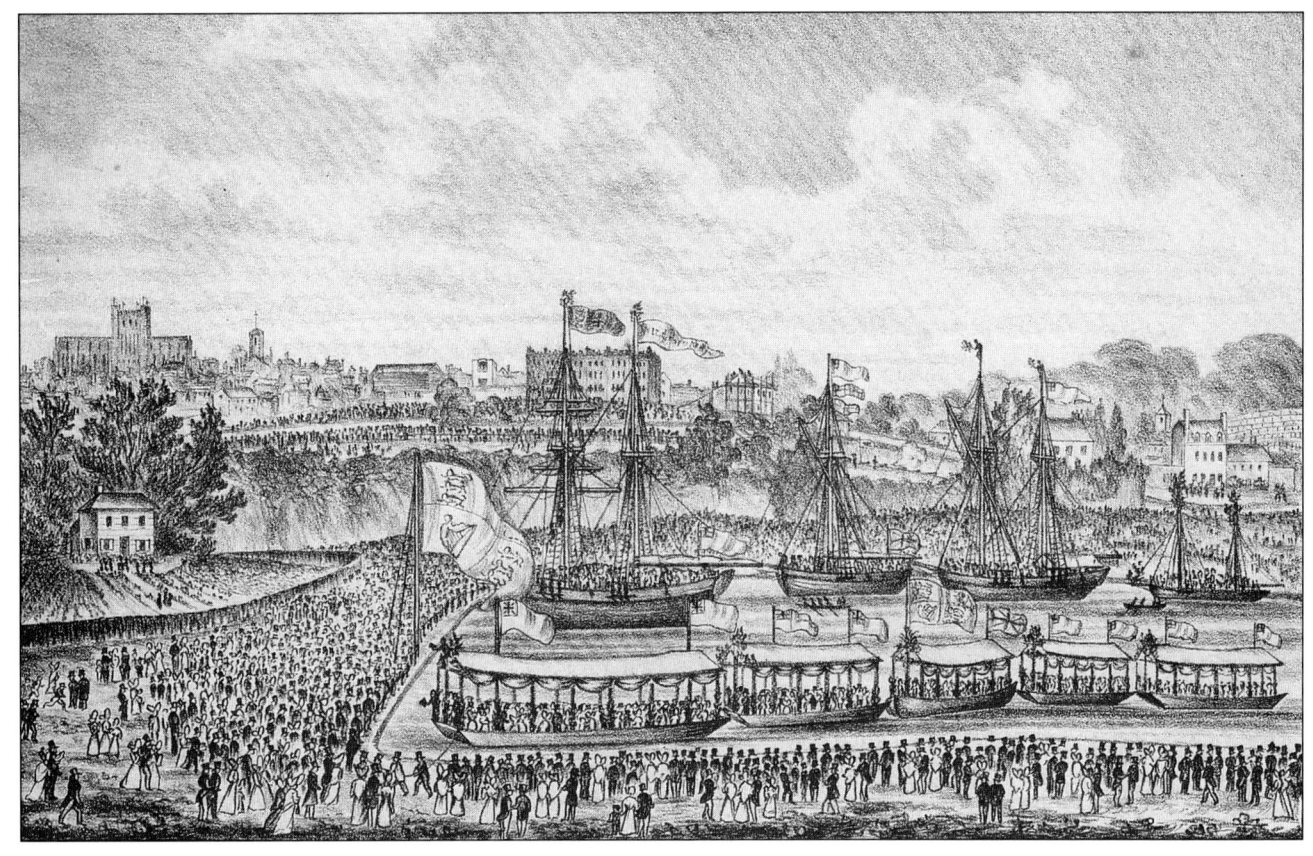

Opening of the New Canal Basin at Exeter 1830

would only allow it to fill with silt. On the other hand a canal of 15 feet instead of 12 feet of water would greatly promote the prosperity of the port.

Work proceeded therefore as Green planned. The canal was extended to Turf and opened on 15 September 1827, the new basin on 29 September 1830 and a new lock at Topsham to replace the facility of the lower sluice to Topsham traders, in September 1832.

So, by persisting, Green first persuaded the city to improve the canal and then raised the specification. When the city sought the guidance of Telford, that gentleman had no hesitation in approving the works.

BRIDGES, CANALS AND TURNPIKES

Aerial view of Exeter Canal Basin photographed in April 1997

However, costs had been very high and, moreover, payments of nearly £90,000 had been made by the treasurer to James Green without authority either for the one party to pay or the other to draw these amounts. Exeter had a fine canal but its finances were to be in trouble thereafter.[7]

3.32 THE BUDE CANAL

The engineering works necessary to enable this canal to develop successfully were substantial.[8] At Bude, a sea lock had to be built and this lock was improved in 1835 by J.M. Rendel to be 116 feet long and 29 feet 6 inches wide in order to admit 300 ton ships to the canal basin. The lock is still in use as the wharf provides winter berths for local vessels. The approach to the lock is protected on the south western side by a stone and rubble breakwater. In February 1838 this was damaged by a great storm and rebuilt by Rendel.

Tamar's Lake earth dam at Alfardisworthy is 180 yards long at the crest and 26 feet high at the lowest point. The reservoir area is 52 acres and the storage capacity is 40 million gallons from a catchment of 4200 acres. Top water level is 441 feet 6 inches above Ordnance Datum. This was an early dam, even for canal works. Nowadays the reservoir is maintained as an area for countryside recreation and wildlife conservation and is accessible to the public at map reference SS295107.

The canal was built to take full size barges from the harbour basin as far as Hele Bridge, 1½ miles upstream, and this length of canal to the bottom of the site of the first inclined plane remains. At the

Tamar Lake – 1993 aerial view with dam (foreground)

time there was an inclined plane to Marhamchurch (120 feet riase) where the narrow canal began and another at Hobbacott Down (225 feet rise) taking the canal to an elevation of about 360 feet. At Red Post the canal branched north to Blagdonmoor near Thornbury and to the reservoir, crossing the river Tamar by the Burmsdon aqueduct before reaching a third inclined plane at Venn (58 feet rise). The southern branch descended the Tamar valley using three inclined planes at Merrafield (60 feet fall), Tamerton (59 feet fall) and Werrington (51 feet fall), to reach the terminal at Crossgate, 3 miles north of Launceston. The narrow canal was 10 feet wide at the bottom, 3 feet deep and 19 feet wide at the top.

The motive power for the inclined planes was water using a bucket-in-the-well system at Hobbacott, supplemented later by a steam engine and water wheels elsewhere.

Bridges, Canals and Turnpikes

Bude Harbour and Sea Lock, 1823

Bude Canal, Hele Bridge (SS214037)

Bude Canal, Keeper's Cottage and inclined plane to Marhamchurch (SS216037)

An interim report on Bude harbour and canal was made to the committee of management on 13 January 1821 by James Green, engineer, stating that the value of the work to be executed was £76,794 and that on 4 January the works executed and materials delivered were to the value of £33,668 while preparation work accounted for £4000. He had received £31,550 on account. He also reported that sand could now be carried to Hele Bridge and that works would be completed to that point by next Lady Day. By August next the canal would be navigable from Bude to Merrafield (9¼ miles) and from Bude to Veala (8¼ miles). By September next the canal would be navigable from Bude to the reservoir (13¼ miles) and altogether 24 miles would be available with considerable progress in the other parts. The report was signed by Stephen Giles, clerk of the company, from the canal office, Dowlsdown, 13 January 1821.[9]

The canal was officially opened on 8 July 1823 to Blagdonmoor and to Tamerton Bridge at a cost of £128,000.[10] The final length was started in 1824 and by 1825 the canal was completed to Crossgates, unable to pass Werrington Park. It remained in use until 1891. Most of its course has now disappeared, but some stretches can be identified, especially with the aid of older maps.

3.33 THE TORRINGTON CANAL

In 1810 James Green had prepared a scheme for a canal from Hallspill to Torrington along the eastern bank of the river Torridge. Presumably he was working for Lord Rolle but nothing came of this. In 1814 Roger Hopkins came from Swansea to Bideford at the call of Lord Rolle and with William Tardrew they planned a tramroad or railway to start two miles from Bideford and to extend towards Great Torrington and the interior of the county. This project also came to nothing. It was in 1823 that Lord Rolle began another canal at his own expense and without parliamentary authority with Green as his engineer.[11]

This time the canal began on the western side of the river Torridge downstream of Weare Gifford Bridge (at SS458234) where there was a ramp with chains into a basin adjacent to the river.[12] After passing the Annery lime kilns at the bridge approach the canal used an inclined plane to reach its final level and followed the new highway to a point opposite Beam Mansion where after a right angle turn it was carried across the river Torridge by a magnificent aqueduct comprising five arches of 31 feet span. It was 21 feet between parapets. Thereafter the canal followed the eastern side of the river passing the lower side of Great Torrington Common to the Town Mills at SS500183.

Lord Rolle also employed Green for building the new grist mill and for the erection of machinery.[13] The new manor mill was erected in castellated form with battlemented gables and parapet walls surrounding an inner yard. 20 years later a new bridge was built across the river nearby by Thomas Whitaker, Green's successor as County Surveyor, and another two arches had to be built over the two terminal branches of the canal that led to the mill.

Today all that is left is the aqueduct converted to carry the approach road to Beam House and the mill buildings with their adjacent arches beneath the highway. In 1989 the mill buildings formed an attractive hotel setting for the Orford Lodge Hotel.

Torrington Canal, Beam aqueduct 1825   (SS473209)

3.34 THE LISKEARD AND LOOE CANAL

In 1823 a committee was formed to make a canal or railroad from Liskeard to Looe in Cornwall to carry inland lime and sand from Looe. James Green was called in and asked to report on the advantages of a turnpike road, a canal or a railroad. He reported that the valley was too steep for a locked canal and recommended the use of inclined planes and that the boats should carry 4 tons of cargo.

At September meetings in Looe and Liskeard the report was considered. The promoters decided to seek a bill with Green being asked to draw the plan which was deposited in October. He envisaged traffic inwards of limestone and culm for Moorswater, coal, iron, timber and merchandise for Liskeard with agricultural produce outwards to the sea.

Plan

of Proposed Improvements in the RIVER TEIGN, and a CANAL from that River to the Town of NEWTON ABBOTT in the County of Devon.

as Proposed by James Green, Civil Engineer

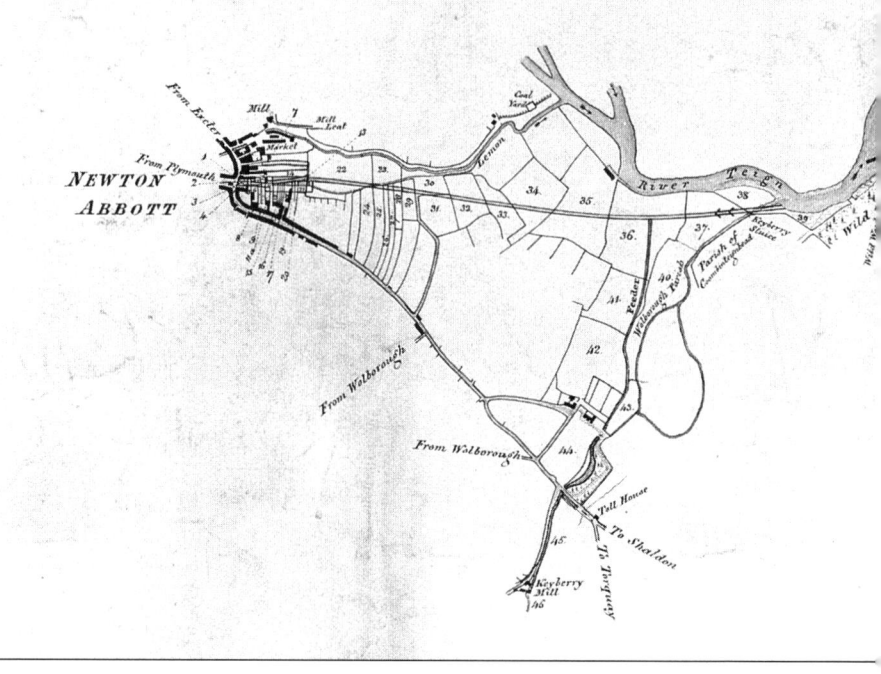

1827

Second thoughts then prevailed, the route was re-surveyed by a Liskeard engineer and an Act was obtained in 1825 for a locked canal with 24 small locks and a larger river lock in nearly 6 miles with a rise of 156 feet.[14] The canal was opened in 1828.

3.35 THE NEWTON ABBOT CANAL PROPOSAL

As a result of a meeting called in 1826 at Newton Abbot, Green was employed in 1827 to consider how shipping in the Teign estuary could be brought to Newton Abbot. He proposed a canal about a mile long with a tide lock to its entrance where the Aller brook discharged to the Teign, running across low ground more or less on the line of the present Queen Street to a basin in the centre of the town in the area now occupied by Courtenay Street.[15]

A plan was deposited but no action followed.[16] However, this was only a few years after the Bude Canal had commenced working and some of Green's ideas prevailed. The Newton Abbot market boats from Teignmouth were fitted with wheels, the bottom of the river Lemon was paved with slabs and cobbles, and horses drew the boats up the paved bed of the river from its outlet to the town centre and also to Bradley mill with hides for the tannery.

3.36 THE WESTERN SHIP CANAL PROPOSAL

The idea of linking the Bristol and English channels had been alive since 1768 when Brindley was

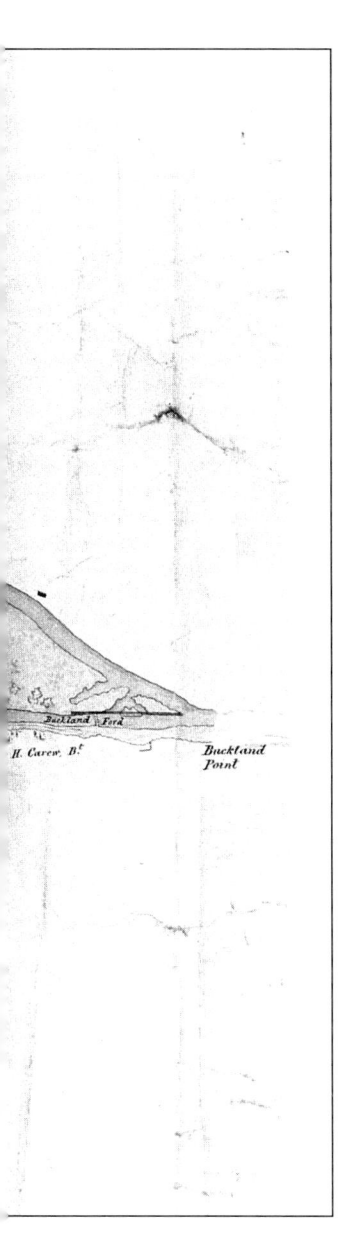

Plan of proposal for canal to Newton Abbot 1827

asked to survey a route for a canal between the channels. In 1810 Rennie was asked to survey a canal and recommended a ship canal for vessels of 120 tons. In 1821 another group of men went back to the idea of a barge canal which would have been more for local use and they employed James Green to make the survey.[17]

Green, obviously with the Bude Canal in mind, suggested a tub-boat canal to run from the Bridgwater to Taunton canal, 2 miles from Taunton, to Beer. There were to be five inclined planes, four tunnels and a pier at Beer to which boats would descend on rails and tip their contents to ships berthed alongside. The cost was estimated at £123,000 and in December 1822 a committee was appointed which included Lord Rolle. A meeting in London supported Green's plan and soon afterwards a Bill was introduced to Parliament.

Others were still supporting the Rennie proposal and by April 1823 the Bill had been withdrawn as opinion was moving towards the ship canal. In June 1824 a meeting in London chaired by Sir T. Lethbridge engaged Thomas Telford, assisted by Captain George Nicholls and James Green, to make a survey which was completed by August with Green signing the plans. They chose a route from Stolford to Beer via Creech St Michael, Ilminster and Chard for a canal 15 feet deep with 30 locks to take vessels of 200 tons.

There was substantial opposition to the passage of the Bill during which Green gave evidence before a Committee of the House of Commons on 24 May 1825[18] but an Act was obtained in July. A shareholders' meeting was held in August and Green began surveying for the work but soon after the scheme was dropped. Nevertheless, much of the survey work must have been useful to Green for the Chard canal.

3.4 THE TURNPIKE ROADS

Having had recommendations for the Exeter to Plymouth road, the Exeter Turnpike Trust quickly turned its attention to the road to Okehampton asking Green to consider its whole length to the limit at Crockernwell. Green rapidly produced proposals for the length from Exeter to Taphouse (Tedburn St Mary) and these were just as rapidly carried out.

In 1765 the main route into Devon from London was via Salisbury, Dorchester and Bridport to Axminster, thence across the rivers Axe and Yarty to Honiton and to Exeter. Travellers journeying via Mere and Ilchester or via Shaftesbury and Yeovil joined the line of the Fosse Way finding their way down the Axe valley to Axminster before turning towards Honiton.

After 1776 the Chard Turnpike Trust established a route via Stockland to Honiton but this was very hilly. After 1808 the Honiton and Ilminster Trust established a direct route from Ilminster to Honiton at gradients acceptable to horse drawn coaches. By an Act of 1817 the Chard Turnpike Trust diverted their Stockland route to cross the river Yarty below Yarcombe to join the Honiton to Ilminster road at a place called Devonshire Inn. Green was consulted by both these trusts to try to find improvements to their routes. Of the three main routes, the Honiton and Ilminster route established itself as the fastest to London[20] and today it is the trunk road from Exeter to London.

As County Bridge Surveyor, Green must have kept in close touch with the various turnpike trusts because in Devon almost all new roads had a bridging implication and eventually the trustees would wish the county to take responsibility for their new bridges. Green would have wanted to be satisfied that the bridges were well founded because of the active rivers, and it is likely that he would have also have wanted to provide the turnpikes with the drawings and be paid a fee for so doing!

His previous relationship with the Plymouth Eastern Turnpike Trust led to his scheme of 1823/24 for a new direct route from Rendel's proposed Laira Bridge through Yealmpton and Ermington and up the Ludbrook valley to Lady Down beyond Ugbrook. This also provided him with the opportunity of building new bridges over the rivers Yealm, Erme and Harbourne.

In 1824 Green built a new bridge over the river Taw at Eggesford for the Hon. Newton Fellowes at the latter's expense. Fellowes was a trustee of the Exeter Turnpike Trust and quickly persuaded them to build a new road to Barnstaple down the valley of the Taw to Eggesford (close to his home) and also persuaded the Exeter Trust to obtain the agreement of the Barnstaple Trust to build a complementary road from Eggesford along the Taw valley to Fishleigh where it joined the existing road to Barnstaple. The benefit of the new route was profound, some 1100 feet of unnecessary ascents and descents between Crediton and Colleton Mills (SS665156) being avoided.[21] Green must surely have put this idea to Fellowes who would have benefited so much in his journeys to Exeter.

Of course, not all of Green's proposals were taken up and the Lyme Regis diversion and the Topsham to Exmouth route were examples, but the totality of his involvement shows that he was respected as an innovative and effective highway engineer.

3.41 POCOMBE BRIDGE TO TEDBURN St MARY

Green was invited by the Exeter Turnpike trustees to make a survey of the turnpike road between Exeter and Crockernwell, the limit of the trust's route, with a view to ascertaining whether improvements could be made and his written report was dated 30 December 1820. When Green made his verbal report to the Trust at the Globe Tavern on 16 January 1821 instead of the usual handful of people, such was the anticipation that 52 trustees were present, including three baronets, Acland, Carew and Northcote.[22, 23, 24]

This road ascended to a height of 750 feet at Heath Cross before descending to Tedburn St Mary and then climbing again to Crockernwell. In Exeter it began with an ascent of 1 in 8 from the turnpike gate to Red Hill and in many other places the gradient was 1 in 9. Green found it difficult to make any suggestions for improvement of the inclines though there were many places where the route could be widened. He was particularly critical of a descent of 1 in 7 to Lilleybrook before the final hill to Tedburn St Mary.

He therefore proposed an alternative route as far as Tedburn because the line of the existing road from Tedburn to Crockernwell could be satisfactorily improved. His proposal was to use the Moretonhampstead road as far as Pocombe Hill where he would descend gradually to Lower Wheatley Farm and then follow the Alphin brook, ascending gently towards Tedburn to a summit of 440 feet near the present Pathfinder village. Generally his gradients were restricted to 1 in 20, which was considered reasonable in those days. At Great Huish Farm an underpass for farm traffic was formed where the new road crossed a small valley, another instance of the early use of grade separation in Devon.

Green's estimate for this diversion of some six miles was £6750 plus £990 for improvements from Tedburn to Crockernwell. The new route was half a mile longer than the old, but saved unnecessary ascents and descents of 375 feet. A recital of the summit heights of the old and new roads does not tell the whole story of the amount of climbing required to follow these routes. The difference in elevation between the Seven Stars Inn at St Thomas and Tedburn St Mary was about 390 feet, but this was not a continuous rise and the table below shows the amount of the three ascents in that direction.

Travelling in the opposite direction in order the descend the 390 feet from Tedburn St Mary to St Thomas, one does in fact have to ascend 570 feet on the Whitestone route and 195 feet on the new

(Heights above Ordnance datum in feet)

VIA WHITESTONE			VIA FIVE MILE HILL		
St Thomas	30		St Thomas	30	
		+271			+170
Redhills (W'stoneX)	301		Redhills (Crossmead)	200	
Nadderwater	200		Pocombe Bridge	95	
		+550			+345
Heath Cross	750		Five Mile Hill	440	
Lilleybrook	281		Great Huish	350	
		+139			+70
Tedburn St Mary	420		Tedburn St Mary	420	
TOTAL RISES		+960			+585

Difference in summit heights 750–440+310
Difference in total ascents 960–585 = 375

Bridges, Canals and Turnpikes

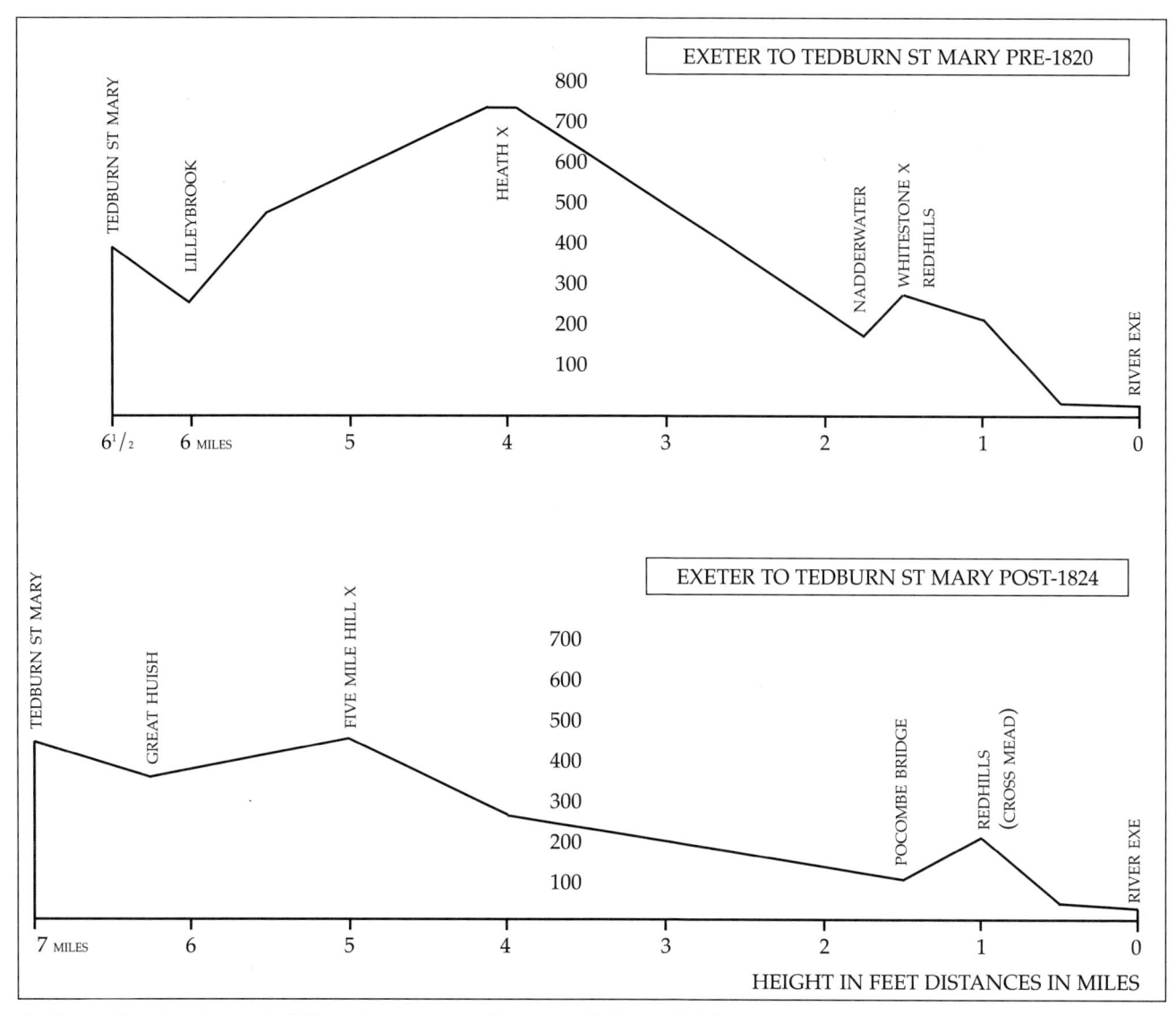

Sections showing impact of Green's new route Exeter to Tedburn St Mary

87

route. This 375 feet difference is the significant measure of the improvement of Green's route coupled with his easier gradients. Braking horse-drawn vehicles on down gradients could be very hazardous so reducing descents was an added bonus.

The new road was opened on 24 October 1824.[25] William McAdam (John Loudon's son) was responsible for the construction work. The estimate proved to be optimistic as McAdam reported in June 1824 that work had cost £6100 to date and that a further £6575 was required to complete. However, both McAdam and Thomas H. Lakeman had checked the estimate and Green was not criticised for this over-expenditure.

Green's fees for making plans for the new road were £185 0s 2d[26] and for surveying, 11 guineas. While his choice of route now seems an obvious one, parts of the road have been costly to maintain in recent years as Five Mile Hill and the length by West Town have been constantly on the move due to local instability of the hillside surface.

3.42 COUNTESS WEAR COMMITTEE OF THE EXETER TURNPIKE TRUST,

On 10 August 1821 the Countess Wear committee of the Exeter Turnpike Trust considered a report from Mr Green for making a new drawbridge across the canal and for improving the approaches or widening or building a new stone bridge over the Alphington brook whose direct course to the river had been cut off by the construction of the canal in 1566. The committee ordered this work to be carried out including the new stone bridge according to the plan produced by Green and under his superintendence at an expense not exceeding £800.

In October 1823 Green, having carried out the work, wrote to the committee asking them to approve disbursements, including sums not paid, amounting to £1100 'exclusive of his trouble'. He acknowledged receipt of £800 from the Chairman, Charles Collyns. Having examined the additional items to be paid which amounted to £350, including an item of £60 for Green's plans and superintendence, the committee approved an additional expenditure of £150, one item being rejected entirely, and Green's fee reduced to £30.[27]

In order to secure the safety of the canal banks against a repetition of the 1824 storms, Green had advised the city chamber of trade to raise the banks 3 feet from double lock to Turf. On 10 April 1830, because the level of the water in the canal had been thus raised, Green reported to the trustees that if an efficient bridge were, in consequence, to be built over the canal it must be of iron and have a bridge keeper's house attached to it and it would cost the chamber of trade £150 to raise the existing bridge. The trustees therefore wrote to the chamber of trade asking them to raise the masonry at the site at their expense.

In December 1830 Green had prepared a plan for a cast iron swivel bridge and estimated the cost at £759, the house to cost an additional £50. A mixed cast iron and timber bridge would be cheaper at £595.

By May 1831 Green had concluded that it was practicable to make the bridge open on one side instead of two and as this would not affect the contract between the trustees and the chamber of trade as to the contribution of their expenditure, he wished to know whether there was any objection on the part of the trustees. No record was made of the trustees' reply.

3.43 PROPOSED ROAD NEAR LYME REGIS

The route from Axminster to Charmouth rises from near sea level to 690 feet. As followed by the mail coaches it is little altered today except for the short diversion away from Penn Cross and through the tunnel built in 1831. In 1822 Green produced a map of a route from Hunter's Inn (near the summit of the present day trunk road) which descended via Uplyme to pass close to Lyme Regis before reaching Charmouth.[28] No part of this route was built, possibly because Lyme Regis was not of sufficient importance to offset the additional distance for through traffic.

On the map is written Green's reasoning for the adoption of his route.

> NOTA. The inconvenience of the hills on the present road from Axminster to Charmouth on the Great Western Road from London is well known, particularly the one descending into Charmouth which is more than half a mile in length, with an average declivity of four inches in the yard lineal, and some parts of this hill are much more steep. The other part of this stage is wholly composed of sudden and deep ascents and descents and withall, extremely narrow, with sudden and dangerous turnings. The improvement of this road is an object of great importance, and most essential advantages would be derived by adopting such a line of road as is delineated on this map, which would have the effect of lessening the hills on the Great Turnpike Road and at the same time open a communication with the sea port of Lyme Regis ...
>
> Between Lyme Regis and the proposed junction of the road with the present line of road to Axminster, it will only be necessary to rise at 2 inches and a quarter per yard for a distance of 12 chains, and the remained of the road had a gradual elevation of not more than 2 inches in the yard on any other part of it.

So 1 in 18 or 1 in 16 were the desired gradients for the traffic of the times.

3.44 FLYING BRIDGE OVER THE RIVER PLYM TO LADY DOWN NEAR UGBOROUGH

On 2 May 1823 an Act of Parliament was passed to establish the need and authority for Laira Bridge, Plymouth,[29] while on 17 June 1823 another Act was passed to provide a road 400 yards from this bridge through Brixton to Modbury with a branch at Addiston Hill to reach the Bittaford to Totnes road at Ladydown.[30] The Ordnance Survey map of 1809 showed a road beginning in Linketty Lane, Plympton, leading to Brixton and to Modbury, and the another road from Bittaford near Ivybridge leading to Totnes.

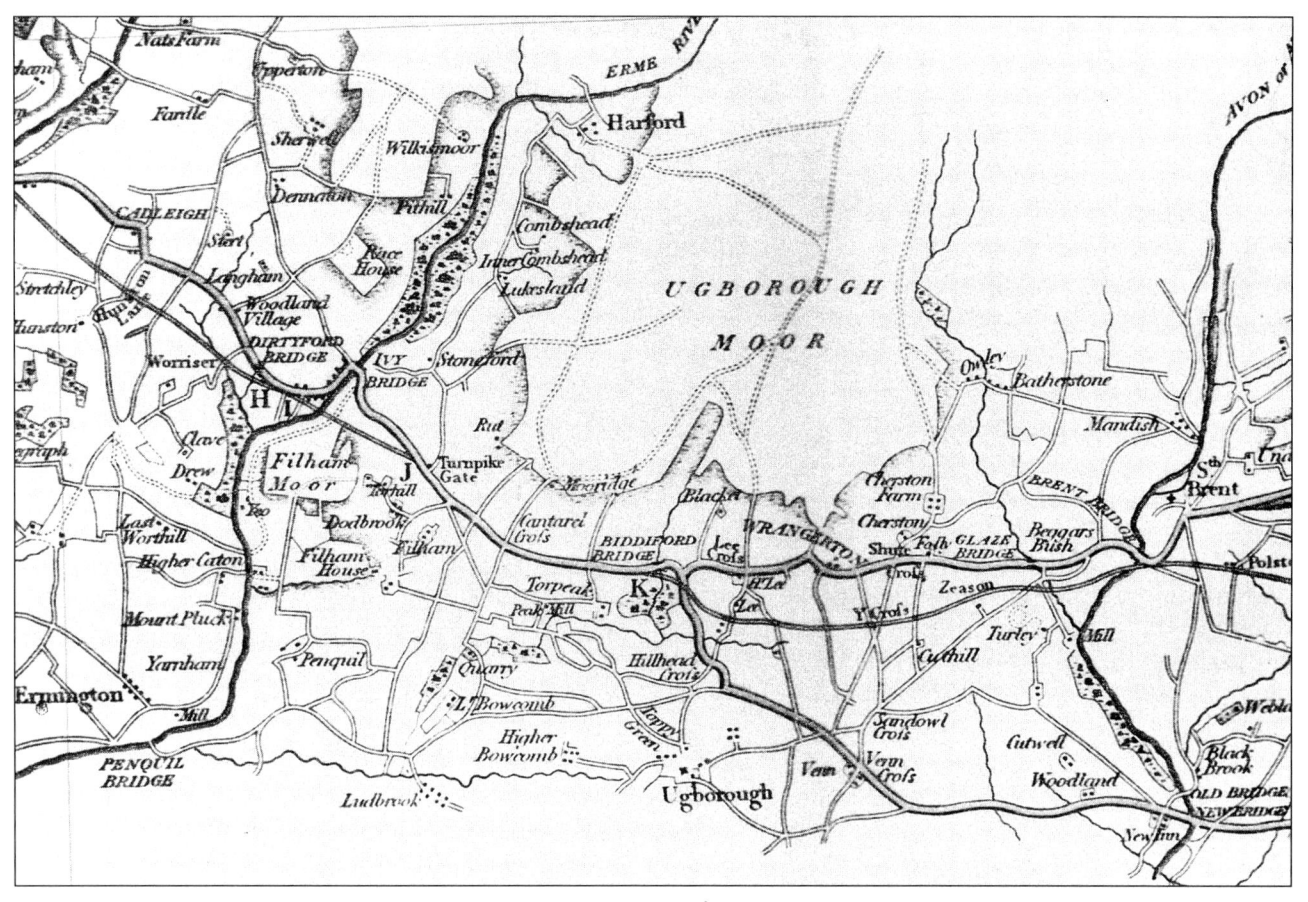

Extract from map of the turnpike road from Plymouth to Exeter, with improvements proposed by James Green, 1819, showing junction with the road to Totnes at Bittaford where it started towards New Bridge at Avonwick

Bridges, Canals and Turnpikes

Lord Morley (Lord Boringdon) had, in 1807, established the Flying Bridge across the river Plym at Laira, so called because it was a manually operated chain ferry for wagons, carts and carriages. Various routes from the ferry via Plymstock and Elberton to Brixton had then developed and on 7 April 1823 a plan was deposited by James Green with the Clerk of the Peace of Devon.[31] Surveyed by Richard Andrews, the scale was 5 inches to the mile and covered a length of 14 miles 4 furlongs and 7 chains. It suggested improvements to a line of road now shown on the Ordnance Survey as part of the A379 and B3210 (A3121); that is, from the Laira estuary diverting through Ermington towards the junction with the Bittaford to Totnes road at the place now called Ladywell at SX694567.

This was a timely survey for the Plymouth Turnpike trustees for in June 1827 James Meadows Rendel would replace the Flying Bridge with his fine, cast iron, five-arched Laira Bridge, so opening up communications with the South Hams on a much improved basis.

Beginning 400 yards from the bridge site, the first deviation from the road through Plymstock suggested by Green was what is now the line of road from Pomphlett Mill presently bypassing Plymstock to Elburton as a dual carriageway.

Alternatives through Brixton and south of Brixton were shown but the line of road through Yealmpton was maintained with a short diversion south of the then Yealm Bridge to eliminate a skew crossing. A diversion south of the old alignment near Way Farm which now exists from SX610520 to SX620520 was shown and next a diversion from SX627520 to SX632524 began what is now the commencement of the road B3210 (A3121) to Ermington. From SX637528 to SX661543 a new line took the traveller across the river Erme and up the valley to Ludbrook. The existing road was then followed to make a junction short of Ugbrook at SX674553, bypassing the village to again join an existing road at SX679557 and continue to Lady Down (Ladywell).

How quickly the work was done is unclear but Green's proposals were closely followed by the turnpike trustees. He provided plans and specifications for the new Yealm Bridge and the bridge over the river Erme and its tributary from Ludbrook, as the county court agreed to share their cost with the trustees (Epiphany 1824). The lengths of new road built were 4000 yards to bypass Plymstock, 1000 yards at Way Farm, 500 yards at the junction of A379 and B3210, 3000 yards from Ermington to Ludbrook and 600 yards bypassing Ugbrook (just over five miles). The other purpose of the deposited plan and the subsequent Act would have been to allow the turnpike trustees to widen the whole route to a standard width and also to deviate 100 yards either side of the proposed line from beginning to end.

Green's diversion, with a maximum gradient from the river Erme to past Ugbrook of 1 in 21, once again using a stream valley, has produced a gradient for the road below his design value of a maximum of about 1 in 16 that he used in his previous schemes for the Exeter Turnpike Trust. The importance of the new road up the valley to

Ludbrook is emphasised by its recent upgrading by the Department of Transport to A3121.

From Lady Down the road had always continued to the place we now know as Avonwick where New Bridge, South Brent (SX715582) had been rebuilt by Green for the county over the river Avon in 1821. From this bridge a road continued to Totnes but in 1825 a three-mile diversion was made just north of Sandwell to run directly to the town via Follaton and Green was asked to supervise the construction of a new bridge over the river Harbourne at SX757598 (Epiphany 1825) to replace Yeo Bridge, the county bridge on the old route.

The importance of this additional length was that a high standard turnpike road with easy gradients was now available from Laira Bridge, Plymouth, directly to Totnes without the traveller having to use part of the Plymouth to Exeter road, an option shown on the 1809 Ordnance Survey, or else having to travel via Modbury on the Plymouth to Kingsbridge road.

The construction of a fine three-arch town bridge over the river Dart in Totnes by Charles Fowler in 1828 removed all river crossing difficulties between Laira Bridge and Torbay.

Today this route from Laira to Totnes is very suitable for the private motorist travelling from Plymouth to Torbay but, once known as the B3210, it is now not clearly marked because of the wish of the highway authority to persuade heavy goods vehicles to take other routes.

3.45 ROADS AROUND STRATTON

In a document marked 'To be returned TDA' (Sir Thomas Acland), Green was requested to examine and report on the various existing lines of road between Stratton and Launceston, Stratton, Holsworthy and Hatherleigh and Stratton and the Bideford turnpike, with a view to pointing out the best lines to be adopted in creating a trust for the formation of turnpike roads between these places with lateral communication to the town and harbour of Bude.[32]

He was also asked to bear in mind 'the expediency of using as far as possible the present parish roads as well as reducing the hills and other obstacles and like wise to consider the commonsense and accommodation of the gentry and general population of the adjoining country and the facility of opening communications hereafter to other principal towns in the neighbourhood.'

Green wrote to Sir Thomas on the 14 July 1824.

> I feel particularly obliged by your sending me the altered Standing Orders of the House of Commons which did not however arrive in time for the paper of tomorrow but (I shall) take care they appear in Trewman's next paper, I conclude the form of instructions to be given to me relative to the survey of the roads has your approbation, if so I must cheerfully accede to them, and I have no doubt that all you wish and that appears to be desired by those instructions can be accomplished for the one hundred guineas you name. I hope however you will be kind enough to extend the

limited time to 20th October which I think may be allowed without injury to the objects of the concern as the ( ) of the House now stand. If I had no other reason, the trouble you have taken to secure me in this business would insure my not neglecting it, but I have so much to do for the next session of Parliament independently of a more than usual brief of this business that any extension of time that you can give me would be a great friendship. The only member of the Parliament I have seen is J. Kennaway Esq who this morning thought the Committee must meet soon but that it was subject to your call. I did not return any answer by your (Messenger) wishes to use that gentleman before I did so but I have not been able to find him. Should you be in town on Friday perhaps there be found members enough to form a Committee. I shall be at home and most happy to attend you.

I am, Dear Sir
Your most faithful and obliged

From this it can be seen that Green's work for the Exeter Turnpike Trust on the Haldon and Taphouse improvements, together with opening up Bude harbour, had caused Sir Thomas Acland and others to decide to make general improvements to the roads inland from Bude and Stratton.

3.46 THE ILMINSTER TURNPIKE TRUST

The Honiton and Ilminster Turnpike Trust had completed their new road, now the A303, between Yard Farm, Upottery, and Horton, Ilminster, in 1812 and this enabled a new coach service between London and Exeter to complete the journey in 26 hours.

In October 1825 the Ilminster Trust (a separate trust) asked the trustees of the Wincanton, Honiton and Ilminster, and the Yeovil trusts if they would employ an eminent engineer to survey the principal lines of road in their districts and report the improvements which might be made, the expense of the survey to be in proportion to the length of road surveyed.

In November, Mr Hanning, Mr Edmonds and the clerk of the Ilminster Trust were instructed to meet the deputies from the trustees of the Wincanton, Honiton and Ilminster and the Ilchester roads, at Ilchester on 22 November and at that meeting it was agreed that Mr Green of Exeter should be employed at joint expense. It should be noted that Mr Hanning was one of the subscribers to the new coach service of 1812 and was also interested in the drainage of Westmoor for which Green would be the engineer.

In July 1826 Mr Hanning drew £50 from the treasurer towards the expense of Mr Green's survey. In July 1827 the trustees wrote to Mr Green disapproving his action in passing through cornfields in June and 'meadows hained up for mowing'. In June 1828 Messrs Hanning, Edmonds and the clerk were requested to meet at Ilchester regarding the survey and in September a further £75 was authorised for Mr Hanning to pay Mr Green.[33]

The confirmation of the route from Yard Farm, Upottery, to the far end of the Wincanton Trust at

Willoughby Hedge settled the alignment of the present trunk road A303 for 150 years.

3.47 THE CHARD TURNPIKE TRUST

By an Act of 1776 the Chard Turnpike Trust established a road between Honiton and Chard via Stockland. In January 1812 the trustees persuaded the Devon Quarter Sessions that Long Bridge, Membury, and Cotleigh Bridge should be repaired at the county's expense and Green carried out this work as County Surveyor. Telford surveyed and reported on the Exeter road between Shaftesbury and Honiton (the present A30) but having decided that Telford's proposals for improvement were too ambitious for their limited funds, the Chard trustees then considered alternative, cheaper, schemes. After 1817 the road from Chard to Stockland was diverted to cross the river Yarty at Yarcombe and thence to join the Honiton and Ilminster road at Devonshire Inn, near Upottery.

In July 1827 the trustees of the Chard Turnpike roads held a meeting to consider the best means of preserving the continuance of the auxiliary mail coach on the present line of the road. They adopted a petition to be presented to the postmaster-general and instructed their clerk to write to Earl Poulett and the clerks of the other turnpike trustees to gain their support. The following month a report of their surveyor, Mr William Summers, was sent to the post office with an assurance that the road surface would immediately be put into the best state.

During August and September Mr Summers was instructed to investigate various lines of a route between Yarcombe and Chard and as these were developed it was resolved that for the satisfaction of the post office and to gain confidence in the proposed new road, Mr Green be applied to for his opinion on the line proposed by Mr Summers and that Mr Green be requested to attend a meeting scheduled for 4 October.

When Mr Green attended, he was asked to report on the best line of road between Chard and Yarcombe keeping in view the lines proposed by Mr Summers. By 6 November Green's report on the alterations and improvements in the road to Yarcombe was approved, a further £1600 above the sum already provided would have to be raised and Mr Summers was instructed to prepare the necessary maps, sections and estimates of the new road and alterations.

On 21 November it was decided to adopt the plans that had been prepared and to advertise for tenders. On 30 November 1827 Mr William Summers deposited a plan with the Devon Quarter Sessions,[34] and on 19 August 1828 the trustees ordered that Mr Green's bill of £51 3s be paid.[35]

The continual improvement of the highways in the coaching era meant that faster routes into Devon came into being very rapidly and by 1836 there were three London mail coaches passing through Honiton daily to and from Exeter. They all left the GPO at St Martin le Grand at 8 p.m. and the first arrived at Honiton via Andover and Ilminster, (154 miles) at 11 a.m. the next day, the second via Andover and Chard, (156 miles) at 12.30 p.m., while the third, via Andover, Dorchester and Axminster, (160 miles) arrived at 1.20 p.m.[20]

3.48 BRIDGES ON THE NEW EXETER TO BARNSTAPLE ROAD

During some twenty years of intense activity the Devonshire turnpike trusts busied themselves re-routing their roads along the valleys to escape the steep ascents and descents to and from the high ground across the county, occupied by the ancient routes. Just as the Exeter to Plymouth and the Exeter to Okehampton roads received this treatment so the trustees of the Exeter and the Barnstaple turnpikes considered the road joining their towns.

Hitherto the traveller from Exeter to Barnstaple would have journeyed via Newton St Cyres to Crediton and then followed what is now a minor road to Morchard Bishop, Chawleigh and Chulmleigh. At Chulmleigh he would have had to make a decision. One route would have taken him to cross the river Mole at Head Bridge and thence via Chittlehamholt, Chittlehampton and east of Codden Hill to Barnstaple, entering the town from the south-east. The other route would have taken him across the river Taw at Colleton Mills to Burrington and thence to High Bickington, Atherington, New Bridge, Tawstock, and Bishop's Tawton, again to enter the town from the south-east.[36]

In July 1825 the Exeter trustees communicated with the Barnstaple trustees regarding their need to improve their road and Mr Pascoe, a former assistant to Green at Braunton Marsh and now surveyor for new works for the Exeter Turnpike, was employed to survey and report, which he did in September. The following March an important Act was obtained allowing the Exeter trustees to make new or diverted roads from Crediton to Barnstaple Cross, from Barnstaple Cross to Copplestone and a new road from Copplestone to Eggesford Bridge to join the intended new road from Barnstaple. In the early part of 1831 the new road to Eggesford was opened.[37, 38, 39]

Three new bridges were required for the new road from Copplestone to Eggesford and an advertisement for contractors was placed in the *Exeter Flying Post* by Mark Kennaway, the clerk to the turnpike trustees at Cathedral Yard, Exeter. The bridges were over the river Yeo at Bury, over the mill leat near Bury Farm and over the river Yeo near Lapford.[40] In due course the Exeter trustees ordered James Green to be paid £60, his charge for plans and specifications of bridges on the Eggesford road.

On the Barnstaple length a large single span was required over the river Little Dart, and not a mile from Head Bridge a substantial bridge over the river Taw was needed at Newnham. A specification for the Barnstaple Trust's new road from Eggesford to communicate with the turnpike road at Fishleigh in the parish of Tawstock was issued by the Trust's surveyor, Charles Bailey on 25 November 1828. It contained drawings for the road in particular places and for the two bridges required, one over the river Little Dart and the other at Newnham, near the present King's Nympton station.[41]

James Green – Canal Builder and County Surveyor

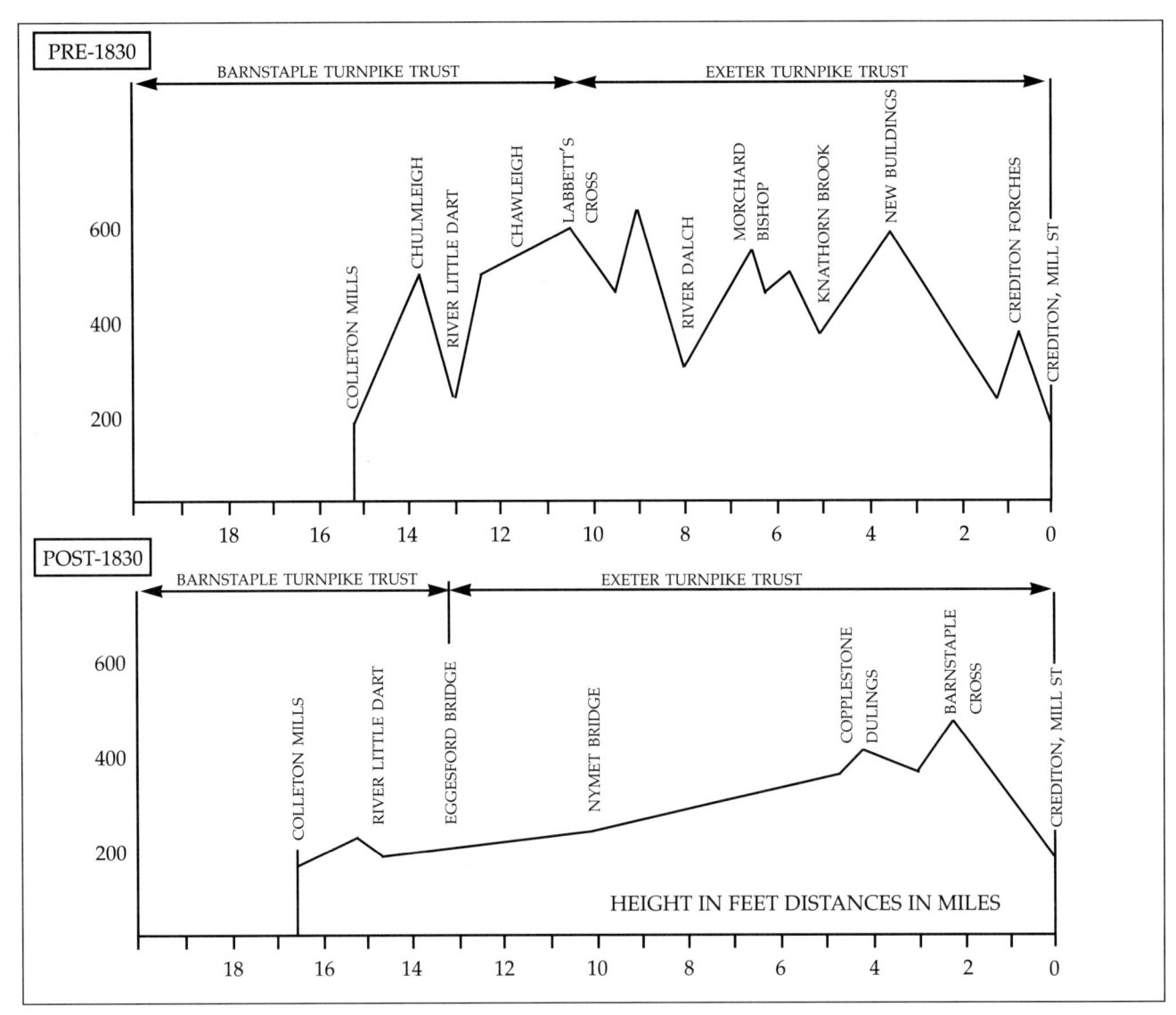

Sections showing impact of Green's new route Crediton to Colleton Mills

BRIDGES, CANALS AND TURNPIKES

A sheet of accounts for the Eggesford New Road, as the Barnstaple trustees called it, listed various payments among which were the following:

    Mr Bailey, for laying down the line of road, estimates etc.   £200
    Mr Green, for plans for Newnham and Dart Bridges   £60
    Messrs Hunkins, for building the Newnham and Dart Bridges   £2000
    Solicitors bill for contracts for bridges, owners of land   £17 5s 2d

The total cost of all the works was £14,055 10s 9d.

So Green prepared the plans and undoubtedly supervised the works because these bridges would become county bridges. What is noticeable is that the contract drawings were different from the bridges as built. Newnham eventually had three arches instead of two, while Little Dart had an arch that sprang from a greater height. Perhaps the most interesting feature of the contract drawings was that the arches were relieved of earth filling by having three internal longitudinal walls to carry transverse jack arches just below the road level, a feature adoped by Telford in his Dunkeld Bridge and his Dean Bridge, Edinburgh. Green's Newnham Bridge followed a style he developed elsewhere of building three segmental low rise arches where the springings are well clear of flood levels.

Whereas the drawings showed the span of the Little Dart Bridge to be 40 feet with a rise of 9 feet, the bridge as constructed has a span of 51 feet and a rise of 8 feet. The Newnham Bridge drawings were of two spans of 40 feet each whereas the bridge as constructed has three spans of 33 feet with a rise of 6 feet in each. Before construction commenced Green probably decided more waterway was justified and in both bridges added 25 per cent to the span dimensions. Both bridges are 20 feet between parapets as shown on the drawings.[42]

3.49 PROPOSED HIGHWAY, TOPSHAM TO EXMOUTH

This proposed route led from Exeter Road, Topsham, through Fore Street and the Strand and directly across the mouth of the river Clyst estuary to Exton. It then followed the edge of the Exe estuary through Lympstone and again along the edge of the estuary to enter Exmouth parallel to what is now Carter Avenue and thence into Exeter Road, Exmouth, near its junction with the relief road.

Much of the proposed route was the one that is now followed by the railway along the edge of the Exe estuary and dating from when the railway was opened in May 1861.

The most ambitious highway work would have been the approach embankments and the bridge across the estuary of the river Clyst. Green anticipated that his route would save one mile from the six and three-quarter mile journey it replaced. He deposited a plan with the Clerk of the Peace on 30 November 1830.[43]

3.5 EARTHWORKS

3.51 BLACHFORD PARK, CORNWOOD

Blachford Park, probably sixteenth century in origin, was remodelled in both the eighteenth and nineteenth centuries. It has landscaped grounds and the lake created in 1827 by James Green was another scheme being carried out from Exeter.[44]

3.6 HARBOURS

3.61 SURVEY OF ST IVES AND ILFRACOMBE HARBOURS

In the years 1823–24, in conjunction with Joseph Whidbey, Green surveyed and reported on the harbours of St Ives and Ilfracombe.[45] Green would have met Whidbey in 1806 when Whidbey and Rennie were working on the Plymouth Breakwater, Whidbey as the Admiralty superintendent and Rennie as the consulting engineer. It was with Whidbey, Rennie and others that Green put his name to the letter of the agent of the solicitor of the Admiralty at Portsmouth mentioned in chapter 2.53.

Joseph Whidbey was a naval officer who had carried out coastal surveys while accompanying Captain George Vancouver on his voyage around the world, and held an appointment as master attendant of Woolwich dockyard. Whidbey was elected in 1809 to the exclusive Society of Civil Engineers, founded by John Smeaton in 1793 (and called the Smeatonians) of which John Rennie was also a member. The society consisted only of eminent people and in 1823 had seven ordinary and 17 honorary members. Whidbey was also a Fellow of the Royal Society and an FLA. He died, aged 78, in March 1830 and is buried in St James' churchyard at Taunton.

3.62 COMBE MARTIN HARBOUR PROPOSAL

Following instructions from a committee considering a harbour at Combe Martin, Green surveyed the bay on 24 September 1827 and produced a report dated 29 October 1827[46] which was published in November in the *North Devon Journal*.[47]

Having mentioned that Combe Martin was about three miles east of Ilfracombe, Green discussed the 'great advantages' of building a breakwater offshore at Combe Martin to enable vessels to shelter from storms from the north-west to north-east when they were unable to enter Ilfracombe harbour. He considered that the tides at Combe Martin were particularly favourable for entering the sheltered water.

However, having recognised that such a work was on a scale beyond the present vision of the committee, Green proceeded to cost three alternative sizes of harbour having piers across a 60 feet wide entrance. The middle sized harbour would have enclosed 12 acres with an entrance about one foot above the low water mark. The piers would have been built with sound ashlar masonry, of good size and bed on the outer side, and with ashlar masonry, though of smaller size, on the inner walls, backing them up with sound rubble stone masonry. The piers would be 22 feet wide at the top, two feet above spring tides with a substantial parapet.

His estimate for the middle sized harbour was £8286. As an alternative he suggested erecting the piers for an inner and an outer harbour and converting the space within the inner piers into a floating basin with gates to maintain the water level at £19,377. None of the works came to fruition. It was unfortunate that rock level on the sea bed was so high in the inlet that it made it completely inaccessible except at high tides. Green suggested removing all outcropping rocks on the harbour bed so that 2 feet of sand could overlay and protect keels from being ripped.

3.63 BRIDPORT HARBOUR SURVEY

In 1809 John Rennie had been advising the harbour authorities on possible improvements that might be carried out and Nettam Giles had produced a survey for him.

In 1823 Green proposed improvements for Bridport Harbour in Dorset consisting of a basin about 400 feet long ENE to SSW connecting to a walled channel about 700 feet long stretching out to the low water mark at spring tides. At the junction of these works was to be an inward facing double gate to maintain the level of water in the basin. The channel width was approximately 50 feet.

Green's proposals were not adopted, a scheme proposed by Francis Giles being carried out in 1824, after an Act of Parliament (4 Geo IV cap 19) had been obtained. Francis Giles seems to have been working on his own account at Bridport but he had worked for John Rennie, senior, and may also have been employed by the younger Rennies after John's death in 1822.[51, 52]

3.64 CARDIFF DOCKS PROPOSAL

In 1814 the second Marquess of Bute succeeded to his Glamorgan estates and became the owner of extensive mineral property in the valleys, the only outlet to the sea being via the sea lock of the Glamorgan canal which was of very limited size.

Lord Bute, after much discussion with others in the district, resolved to construct a dock and this work was eventually carried out at his sole cost thus starting the development of the docks at Cardiff. It was the beginning of a new era of commerce for the city and amongst the engineers and others consulted were Thomas Telford, David Stuart, William Cubitt and James Green.

The scheme prepared by James Green was the one submitted by Lord Bute to Parliament in 1830 for which an Act was obtained. This provided for a ship canal across the mud flats from the mouth of the river Taff to the East Moor, about one and a half miles in length. It was to be enclosed by walled banks and towing quays with a sea lock with flood gates at the outer end. The water level in the canal was to be maintained above spring tide level by means of a fresh water supply from a feeder so as to keep out the silt laden water from the impounded area so preventing precipitation of the mud and obviating the need for maintenance dredging.
As there were difficulties in getting any contractor to tender for the work on account of the risks involved, on the advice of William Cubitt the scheme was abandoned in favour of an open channel cut through the mud to a basin and lock at the shore and leading to a dock to be excavated out

of the East Moor. The dock was opened in 1839 and became eventually the West Bute Dock. This dock was closed in 1964, was subsequently filled in and the site is now a trading estate.[19]

3.7 RESIDENCES

In 1822 Green was living at Elmfield but sometime before 1826 he sold Elmfield to George Sparkes and moved to 36 Southernhay Place. This house was in the style of the other houses still existing on the western side of Southernhay and then occupied by the foremost Exeter professional people. No. 36 was destroyed by bombing in May 1942.

3.8 THE INSTITUTION OF CIVIL ENGINEERS

The Institution dates from 1818 when, at a meeting of young engineers at Kendal's Coffee House in Fleet Street, Henry Robinson Palmer declared that civil engineers needed a headquarters where information based on recorded experience could be available and views exchanged on the application of scientific principles to civil engineering practice. The young engineers later decided that to boost their new organisation a distinguished engineer was needed as President. Thomas Telford was the obvious choice and through his influence the Institution was granted a royal charter by George IV on 3 June 1828.

Telford established a unique library for the Institution and made it a centre for the accumulation and exchange of engineering knowledge; he gathered new members from all over the world. At home he had been actively increasing membership since he became President in 1820. Having worked with James Green on the Western Ship Canal project, he proposed Green as a corresponding member on 6 April 1824. Corresponding membership was available to members outside the radius of the threepenny post from the Institution's Buckingham Street, Adelphi address in London. Green signed his agreement to abide by the rules of the Institution on 30 April.

Three supporters were needed in order to join the Institution at that time and it is interesting to examine the records of the other two, particularly since they had only just been recruited by Telford themselves.

Robert Sibley was eight years younger than Green and in 1818 he had been elected County Surveyor of Middlesex where he worked on roads and bridges, the enlargement of several prisons and the construction of Hanwell lunatic asylum. In 1828 he was appointed to the district of Clerkenwell and in 1832–33 promoted a system of wharf walls with grooved cast iron piles and facing plates, backed with concrete and these were successfully employed at the Island Leadworks, Limehouse and on the large wharf at each bank of the Thames at London Bridge. Sibley became surveyor to the Worshipful Company of Ironmongers in 1839 and served on its council for many years. He had been proposed by Telford to the Institution on 9 March 1824 and signed his agreement on 25 March. He died on 31 March 1849, just a month after Green.

Thomas Jones of Charing Cross, the third supporter, had been proposed for admission on 2 March 1824 and signed his agreement on 29 March. He was a mathematical instrument maker and withdrew from the Institution on 22 August 1834.

From the beginning of the publication of *Transactions* in 1836 members were encouraged to present papers and James Green did so on three occasions from 1838 onwards, as we shall see later.

3.9 TOWARDS THE NEXT DECADE

As the decade ended, Green ran into trouble with his employers. His salary had been increased dramatically during his service. From £300 a year in 1808, it had been increased to £400 in 1812 and to £550 in 1815. By 1826 he was being paid £550 plus £200 for services rendered, plus £120 for plans plus £10 for obtaining a clerk of works, making £880 in all.

At the Devon county Sessions commencing Tuesday 20 April 1830, a complaint was made that the keeper of the sheriff's ward seemed to have an unconditional power to order what he thought fit – no magistrate visited the prison – and as there was no visiting magistrate no order should be paid unless sanctioned by the magistrates of the division of Wonford. In the matter of repairs there was no signature of Mr Green. And now criticism of Green erupted.[48]

Captain Buller complained that, while Mr Green presented an annual charge of £1200 for repairs of public buildings, this seemed extraordinary as there had been nothing like parsimony during their erection. The ward had been erected in 1821, the Bridewell in 1811 and the gaol in 1794. Though it was continually reported from Sessions that the county buildings were in good repair, yet the annual charge of £1200, a large sum, was still incurred and set down under the head of repairs. Captain Buller might be in error but it did appear to him that there must be a fault somewhere – either the work had not been done properly at first or it had not been attended to properly afterwards. As regards their surveyor, with a salary of £550 per year, Captain Buller feared that he had too much business of his own to attend, as he should, to that of the county.

Green defended himself by saying that nothing had ever been done without the directions of the visiting magistrate and that regarding the erection of such buildings as he had been connected with, he would pay the expense of a survey by the first architect in the kingdom, if he did not, after inspection, say that they were well done. As to the repairs in such extensive erections, they would inevitably arise, but as to the gross annual sum, he imagined it must contain many charges to which the term 'repairs', as understood by the magistrates, were not strictly applicable.

All this revealed a complete lack of a system of accountability by the officers responsible for spending the money voted by the Quarter Sessions and a surprising lack of involvement of the magistrates in the supervision of work on the prison, bearing in mind that there was obviously no system for recording expenditure on a day by day basis or a job by job basis.

Mr Fulford therefore gave notice of two motions:
1. Repairs, in the future, should be by contract, either annually or otherwise.
2. For supplies, contracts should be made at such periods and as often in the year as might be necessary.

It was also proposed that Mr Green's salary should revert to £300 per annum as bridge surveyor and £100 as surveyor of the county buildings, but it was agreed that this would be considered at the Michaelmas Quarter Sessions.

By now Green had left Elmfield and was living at 36 Southernhay Place. He was in great favour with the city of Exeter for his work on the canal and Wednesday 29 September 1830 had seen the opening of the basin, or wet dock, at the head of the canal by the Mayor. Woolmer's *Exeter and Plymouth Gazette* devoted a full page to a description of the event and the works concerned giving credit to the 'arrangement and superintendence of Mr Green and his son' of the procession of boats from double locks to the floating dock. It was noted that the length of the basin was 917 feet and its width 110 feet 6 inches over two thirds of the length and 90 feet at the entrance end. The uniform depth was 18 feet.[49] No doubt this represented 'the business of his own' that concerned Captain Buller.

Green's successful completion of work on the Exeter Canal for the city may well have raised jealousies and drawn attention to his advantageous position with a base salary as County Surveyor combined with permission to undertake professional engagements elsewhere in the county and adjacent counties. The decade 1821–30 must have been very lucrative with money coming in from his work for the turnpike trusts and for the canals at Bude, Torrington and Exeter.

It has to be remembered that when Green was appointed, the committee required the surveyor to inspect every county bridge at least once a year, and we have seen that the magistrates who were ultimately responsible for the bridges wished to be present at the inspections. Here was the fundamental requirement of a successful bridge maintenance system, for the early detection of a structure being undermined by the flow of the river, or the masonry becoming loose because the mortar has deteriorated, enables repairs to be effected before more extensive, expensive works are required.

With the establishment of contracts for the maintenance of the bridge road approaches and their parapets and guard rails, Green was proceeding step by step towards the 1831 contract where the contractor would also be expected to keep general watch over the bridge structure. This assisted in ensuring that the general condition of the bridges would also be watched by someone reporting to the County Surveyor, and would logically have kept the need for his inspections to remain at one a year.

Undoubtedly this had been Green's most creative decade. His county bridge work had proceeded satisfactorily at the reduced pace required by

Quarter Sessions, his canal work had establised his reputation in that field and his projects for the turnpike trusts, though less remunerative than his work on canals, had brought him the admiration of the gentlemen of Devon and West Somerset.

It would have been interesting to have been able to examine his bank balance. His pay from the county was substantial at £550 per annum plus expenses. The turnpike trusts had added to this by £50 to £100 each year and one would have expected the canal remuneration to have peaked in 1824–25 but to have remained significant through the decade. The treasurer of the Exeter Canal paid £3954 to the engineer up to the passing of the 1829 Exeter Canal Act.[50]

By this time Elmfield was in the possession of George Sparkes, another Quaker, who with Thomas and Joseph Sparkes had taken control of the General Bank in Exeter in 1818. Presumably the sale of the Elmfield estate had provided more than enough capital for the purchase of 36 Southernhay Place.

One has to ask, 'Where did all this money go?'

CHAPTER 4
1831-1841
BRIDGES, CANALS AND NEWPORT DOCK

4.1 INTRODUCTION. 1831–1841

Following the successful opening of the Exeter Canal Basin three weeks earlier, the mayor and chamber of Exeter on Wednesday 20 October 1830 voted the Freedom of the City of Exeter to James Green. There were 586 Freemen of Exeter in 1835 out of a population of 28,285 and this recognition was the more significant because Green was a Quaker and in the nineteenth century it was never quite socially respectable to be a Nonconformist or a Liberal in the city.[1,2]

Green's stock in Exeter was high despite (or perhaps because of) the cost of the work he had induced the city to carry out on the canal. The canal was now an efficient outlet to the sea ready for its first steam vessel to work through in September 1840.[3]

By now, with all the drawings that had to be produced to support his county and consulting work Green must have had substantial office support while he was working around and away from the county. One of the Exeter Canal drawings is annotated J. Green jnr. His son, Joseph, entered the roll of the Institution of Civil Engineers in 1832 when it was recorded:

Joseph Green of Exeter, practising as an assistant engineer was proposed as an Associate Member on 14 February 1832.

His proposers were William Bruton, James Walker and William Cubitt and Joseph Green agreed to abide by the Rules on 28 February 1832.

This decade, when Green was aged between 50 and 60 years, encompassed the last period of his 33 years as a County Bridge Surveyor and included extensive canal work in Somerset and South Wales. He must have been desperate to earn as much money as possible privately to offset his reduction in salary from the county which took effect in 1831.

Again Green drew on the breadth of his experience to find private work. In 1832 he made proposals to Torquay for its water supply and sewerage and at the same time put forward a scheme for a railway from Newton Abbot to Torquay. The proposals were deposited with the Clerk of the Peace for Devon. In 1833 he carried out the drainage of a portion of the Somerset Levels at Westmoor near Kingsbury Episcopi, an endeavour that has been acclaimed as a successful scheme to cope with both wet and dry seasons. He even undertook a survey for an alternative route for a London to Birmingham canal which he presented in 1836.

Undoubtedly he spread his attentions too widely for he made serious faults in his design for the inclined plane on the Grand Western Canal and overspent on the Kidwelly and Llanelly Canal. The collapse of a dock wall at Burry Port was the last straw and at the end of January 1836 he was removed as engineer from four projects, the Grand Western Canal, the Chard Canal, the Kidwelly and Llanelly Canal and from Burry Port.

Once again his contractual and supervisory finances seem to have been intermingled for he was adjudged bankrupt in 1837 and moved his home from Magdalen Street to Alphington. No doubt he would have been disowned by the Religious Society of Friends. He allowed his son Joseph, 20 years of age in 1837, to supervise his county bridge work and off he went again to South Wales to take over as engineer to Newport Docks. Without a home in Exeter he began to lose control of his bridge work to the extent that the magistrates noticed his son performing some of his duties and in 1840 Green was called to account at the Midsummer Sessions.

Mr Sillifant complained that Green was renting a house in Wales, that he was in the service of the Newport Dock company and that he had had no house in Exeter since last Christmas. Colonel Fulford responded that at £300 a year Green was not paid enough and Green added that there were now 340 bridges to supervise.

In fact the supervision of 340 bridges may not have produced very much work for Green to do. Once a group of bridges has been properly maintained for two decades there may be few surprises. Green had tended to rebuild bridges with adequate spans so that the flood waters through the arches did not scour the undersides of the foundations. Certainly this appears to be the case with his bridges 155 to 190 years later.

Green was given a year's notice from Midsummer 1840. Not all the magistrates were unsympathetic to him but he was too elusive to be retained as bridge surveyor. He was coming up to 60 years of age and his Freedom of the City of Exeter meant nothing to the county magistrates.[4]

4.2 THE COUNTY BRIDGES AND THE COUNTY SURVEYOR

Many substantial single and three-span bridges were built during this decade and Green built two further bridges with cast iron spans of 50 feet. These cast iron spans are particularly interesting because none of them lasted as long as his masonry work and the reasons for this are discussed later. The decade began with a slight alteration to the administration of receiving tenders that underpinned Green's authority in the county.

Maintenance was further defined and responsibility for inspections delegated to the contractors undertaking maintenance contracts for the road approaches. The number of bridges had increased from 236 in 1809 to 276 in 1831. Green's private work with the Barnstaple Bridge trustees widening this town bridge was very successful and is recorded today on the corner of a nearby building. Overall, his county work, despite absence on private work in adjacent counties, appears to have been thorough and to have continued to improve the bridge stock.[5]

Work on the county buildings was confined to maintenance. Drainage work at the prison became important with cholera in Exeter and so the drain situated at the bottom of the Gaol Fields (the Longbrook) was piped. The first proposals for a county lunatic asylum were estimated while the Secretary of State took over the appointment of the prison governor and Home Office inspectors began reporting on the prison. The Exeter Turnpike Trust was formulating its proposals for New North Road and Green was deputed to discuss the fencing of the Gaol Fields with the trustees.

4.21 DEVON QUARTER SESSIONS ADMINISTRATION

The committee of the Quarter Sessions appointed to report on the salaries of the keeper of the gaol, the County Surveyor and the surgeon did so at Epiphany 1831. It recommended that one annual survey of the county bridges, instead of two, should in future take place and the salary of the County Surveyor should be reduced from £550 to £450 including stationery, postage etc. Mr Hamlyn moved as an amendment that it be reduced to £300 and produced a letter from Mr Rendel, a civil engineer with high testimonials, including those of Lord Morley and Mr Telford, in which he offered to perform all the duties specified for £300 a year. On a division, the amendment for reducing the salary to £300 was carried by a majority of 21 to 9.[6]

Green must have been dismayed that James Meadows Rendel, 20 years his junior, should have set his sights on the job and that he should have the support of Lord Morley, for whom Green had worked in 1808, and of Telford who had so recently been supporting Green when checking the proposals for the extension of the Exeter Canal and the construction of the canal basin. Rendel was a brilliant engineer, as was shown by his construction of Laira Bridge for Lord Morley and his innovative work on the Dartmouth floating bridge, which was then being constructed. Rendel eventually stayed in Devon until 1838 when he moved to London and established himself as a foremost engineer in the design of docks and harbours.[7]

Green's successful completion of work on the Exeter Canal for the city may well have raised jealousies and drawn attention to his advantageous position with a base salary as County Surveyor combined with permission to undertake professional engagements elsewhere in the county and adjacent counties.

At the Quarter Sessions of Michaelmas 1830 Green reported that the contracts for keeping up bridge approach roads and repairing parapets etc. had terminated and that many parishes wanted to employ their poor in this work. He also stated that the trustees of the turnpikes appeared to wish to take the repair of those portions of the road, in such a way as they could agree with the county, into their own hands. He therefore requested the direction of the court upon the matter. Sessions ordered that the bridge committee should be revived and that the whole be referred to them.[8]

In September 1831 James Green issued a specification of works necessary to keep in repair the parapets and railings of the county bridges together with the roads over the bridges. (Appendix C) Contractors were expected to tender for the upkeep of the bridges over a period of five years and to help them be sure which were the bridges in their area of operations. The specification was accompanied by a list of 276 bridges. These were grouped in divisons and against each division was also listed the magistrates who would be responsible for the supervision of works in each division.

After defining the work that would be required for the parapets and railings to keep them in sound condition, the specification declared that the contractor responsible for the repair of a bridge would be expected to keep a watchful eye over the state of the masonry on all other parts of the bridge which were not included in the contract. From time to time when repairs became necessary, the contractor was to give information to the magistrates appointed and to the County Surveyor of the precise nature of the repairs required. He was not to carry out the repairs until he received an order in writing from either the magistrate or the surveyor.

The effect of this instruction was to set up a system of bridge inspection and maintenance that ensured that there was a local person who would examine the bridge regularly for defects. By such action the cost of bridge maintenance could be kept to a minimum on the 'stitch in time saves nine' principle.

The list of bridges for which the county was responsible in 1831 is included as Appendix B. Between 1809 and 1831 the number of county bridges had risen from 236 to 276. Dropped from the 1831 list were New Bridge, Gunnislake (Tavistock), Staverton, a bridge adjoining Axbridge, Holne and Holne New Bridges and Newland Bridge at North Tawton.

In came nearly 50 other bridges. To replace Staverton was Emmetts (otherwise Hood or Riverford) on the present Buckfastleigh to Totnes road. Many new names were of bridges on roads administered by the turnpike trusts. From Barnstaple Turnpike Trust there were Bishops Tawton, Winters (at Chapelton), Hansford, Landcross and Iron Bridge, Landcross. From Exeter Turnpike Trust there were New Shuttern, Downes, Sandford, Calves, Ellerhayes, Padbrooke, Topsham, Ide, Broadclyst and Minchinlake.

Long Bridge at Membury, rebuilt by Green in 1812, and Cotleigh Bridge were also newly listed. They were on the Chard Turnpike Trust route of 1776 to Honiton and were in a detached part of Dorset. They still bear plates today warning potential offenders not to damage the bridge on penalty of transportation for life. Another new bridge was not on a turnpiked road. Thorverton Bridge, built in 1792 and rebuilt by Green in 1813, crossed the river Exe in a single span of 84 feet.

Bampton Bridge was at the end of a Tiverton Trust route and at the commencement of a Minehead route. Bickleigh was on a Tiverton route to Exeter while Dart and Stockleigh were on that Trust's route to Crediton.

Remarkably the year 1831 began with a slight shift in Green's favour in the administrative procedures for the construction of a new bridge for the county. This was the case of Shuttern Bridge in Newton St Cyres for which the advertisement ran as follows:[9]

> To Builders and Masons
> Any person desirous of undertaking to build a new bridge at Newton St Cyres may see the plans and specifications on application to James Green, Civil Engineer, Exeter and may deliver sealed tenders for the same (endorsed tender for Newton Bridge) to the said James Green on or before Saturday 23rd of this instant.
>
> Exeter, April 12, 1831

There is no mention this time of viewing the plans at the house of a local magistrate or a local shopkeeper and no mention of returning the tenders to a local magistrate or to the Clerk of the Peace.

This was a small bridge and at the Epiphany Quarter Sessions of 1831 the magistrates had ordered that no more than £100 was to be laid out provided neighbours raised £100 by subscription. Green obviously had the full confidence of Quarter Sessions in his integrity.[10]

At Easter 1831 Green was formally reappointed County Surveyor for a salary of £300 a year for which he was to perform all the duties hitherto performed by him and to prepare all plans of any alterations or improvements in any of the county buildings already erected or hereafter to be erected. He was to perform all other services that might be requested of him by the court as County Surveyor without being allowed any charge therefore – his extras of the 1820s were no longer to be allowed!

At Epiphany 1832 Colonel Fulford, the chairman of the Sessions, and magistrates acting at the castle were ordered to form a central bridge committee to take such matters relating to the county bridges into consideration as might be expedient, while at Midsummer, the committee of accounts noted that the expense of bridging in the Easter Quarter was £400–£500 less than a year ago but this was counter-balanced by the increase in the number of criminals and hence the cost of conveyance, prosecution and gaol maintenance – an early instance of competition between departments of the county for money![11]

The imposition of tolls on county bridges was considered the following year when the Clerk of the Peace of the county of Buckinghamshire wrote to Devon inviting their co-operation in an application to Parliament for collecting a toll on the county bridges. This idea was declined by the Devon magistrates. At Easter 1833 the County Surveyor reported that county bounds and the roads over bridges in many divisions were not being kept properly in repair according to the contract and this matter was referred to the bridge committee.

At the same time the County Solicitor was given a list of bridges to enquire into the liability of the county for their repair in conjunction with the County Counsel. The bridges were:

Northbrook in Topsham   Farrants in Dunsford

Cranford in Broadclyst
Loxbrook in Broadclyst
Langford in Newton St Cyres
Bray in Chittlehamholt and South Molton.
Monkerton in Pinhoe
Coach and Horses in Honiton Clyst
Thelbridge Ford in Sandford

All except the last resulted from the clerk of the Exeter Turnpike Trust indicting the county (as he had reported to the Trust on 6 February 1833 and including Alphington on his list as well). At Midsummer the solicitor advised the court to plead not guilty to Farrants and Coach and Horses but to take over all the others. At Michaelmas the Solicitor advised that the court was liable to the repair of half of Bray Bridge, the other half to be the responsibility of the borough of South Molton. At the same time it was ordered that Alphington Bridge should be rebuilt. The Exeter Turnpike Trust was keeping up the pressure not only to have the maintenance of its bridges assumed by the county but to have them rebuilt to the county's standard.

Dissatisfaction with the form of the bridge contracts was evidenced when the court ordered the County Solicitor to enquire and report on the contracts then in existence for the repair of the bridges and their roads, and to revise them with the assistance of Mr Hamlyn, a magistrate.

At Epiphany 1835 Mr Hamlyn put down a motion for a committee to enquire into the fees of the Clerk of the Peace, but this was refused. The magistrates were not happy, however, with the notice that they received for examining bridges from the County Surveyor, and Mr Elton proposed a motion for the next Sessions that a longer time than three days should be given by the County Surveyor to magistrates previous to his visiting their bridges. This must have caused inconvenience from time to time, but no more was heard of this matter.[12]

At Midsummer 1836 it was noted that contracts for bridges and roads would expire at Michaelmas and it was ordered that the repair of the roads over the county bridges should be offered to the respective turnpike trusts upon the same terms as the trustees now repaired the contiguous roads. A committee would be appointed to advertise for tenders at the same time and to correct and have printed the necessary list of bridges and specifications.

At Epiphany 1837 the bridge committee reported that they had received tenders for repairing the county bridges and roads for five years from Michaelmas last and had contracted with various persons whose tenders were lowest and most eligible and had directed that sureties be taken from them for the due performance of their contract. There was an increase of about £100 in the total amount of the present contracts over the previous five years, but in that period very many bridges had been added to the list and this would account for the difference.

The committee had been asked to apply to the trustees of the turnpike roads near and adjoining the county bridges to ascertain if they would undertake the repair of the county bridge roads. In some instances the trustees declined, in others they offered to undertake it but they generally exceeded the tenders of private individuals and the committee considered that, though the roads might be contracted for by the trustees, the bridges must themselves be let to other parties and that, on the whole, the most simple and advisable plan was to let the whole of the repairs to private persons.

That stated the court's view on contract quite clearly but by Midsummer the County Surveyor was reporting that contractors had greatly neglected the repairs of the county bridge roads and the court referred the matter to the bridge committee to take action. This led at Michaelmas to the magistrates of the Teignbridge division being ordered to consider whether Henry Couch Creagh had performed his contract properly and, if not, whether to discharge him and use the money in the hand of the county to have the work done. Alternatively they might pursue Creagh and his sureties by legal means to ensure that the necessary work was completed.

By Easter 1838 the magistrates were considering the removal of bridge bound stones and claims of the trustees of the turnpikes that the county was not maintaining the roads to the full length of their 300 feet responsibility. No mention was made in the Quarter Sessions minutes of the bankruptcy of the County Surveyor in 1837.

At Epiphany 1839 the Totnes Turnpike Trust wished to know whether they could build a toll house at New Bridge, South Brent, (Avonwick today) close to the bridge and within the bridge bounds. This was agreed provided the trustees erected the house and gate in the position approved by the County Surveyor and on condition that no carriage passing the gate would have a toll exacted if it were employed in carrying materials for the repair of the bridge or road approaches which were the responsibility of the county.

At Epiphany 1840 the court found it necessary to order that in future in all cases when the county was indicted for non-repair of a bridge the County Solicitor should immediately receive a notice of this from the Clerk of the Peace and proceed to make the necessary inquiries as to the liability of the county so that the result would be reported to the next Quarter Sessions. Obviously the Clerk of the Peace had been dilatory in referring the case of Derridge Bridge, near Morchard Bishop, and had waited until Sessions before raising the matter so delaying a decision for some six months.

At Easter the treasurer was instructed to tighten his control and maintain accounts as far back as could be remembered for the repair and rebuilding of each of the county bridges. These should be placed on the table at every Session. Mr Sillifant, junior, gave notice of a motion for the next Sessions that the County Surveyor should reside in the county.

As a result the court immediately ordered a committee to be appointed to look into the way in which the duties of the County Surveyorship were at present discharged and into any improvement which could be made and to report to the next Sessions. The committee consisted of Colonel Fulford, Mr Sillifant junior, Mr Gordon, Mr Drewe, Mr Kekewich, Sir John Kennaway, Sir John Duckworth, Mr Buller, Mr Hamlyn, Mr W.I. Clark and Mr Praed.

Many of these committee members had known James Green for a very long time through their dealings with him in their capacity as trustees of the Exeter Turnpike Trust.

At Midsummer 1840 the committee made their report which was as follows:
1. That it appeared from Mr Green's statement that the duties of the surveyorship are at present in great measure discharged by his son, he having informed the committee that although he makes a point of visiting, inspecting and preparing the plan and specification for all works which he considers important, yet that those works which he considered of minor importance he had confided to the superintendence of his son.
2. That it is the opinion that the discharge of the duties of surveyorship by deputy to this extent was not contemplated on his appointment and cannot be satisfactory to the county.
3. That in the opinion of the committee the surveyor ought to be required to be entirely resident within the county as well as to discharge personally the duties of the office.
4. That it is advisable that whenever fit opportunity occurs that the survey and superintendence of the county buildings be confided to an architect resident in Exeter and be kept distinct from the survey of the bridges.
5. That the County Surveyor should be required to attend at every Quarter Sessions.
6. That having regard to the fact that the majority of the county bridges are situated within the command of Exeter as a centre and also to the importance of maintaining the responsibility and independence of your surveyor, your committee do not recommend the division of the county into districts more or less numerous for the survey and care of the bridges.
7. That it is advisable that the surveyor do make two annual surveys of the county bridges.
8. That the time occupied by the surveyor in making such surveys would be considerably shortened by a regulation to the following effect regarding the meetings of the surveyor with the magistrates of the several districts at the county bridges and the committee recommends its adoption:

> Surveyor whenever required to do so by one or more neighbouring magistrates to fix the time of his attendance at any bridge or bridges which such magistrates may name and to give such magistrates at least ( ) days previous notice thereof.
> Surveyor also to give like notice to magistrates in special cases in which he may desire their attendance and directions.
> In all other cases the Surveyor not to be required to give notice to the magistrates of the time at which he is coming or to await their attendance.
> By order of the Committee
>
> (Signed) J. Kennaway

This was confirmed by the court at Michaelmas 1840 when it was resolved that separate salaries be fixed for the superintendence of the bridges and the county buildings and that both offices should not necessarily be held by the same person; also that the court should proceed to the election of a surveyor at the next Easter Sessions. Notice had been given to the County Surveyor at the

Midsummer Sessions 1840 that the county would terminate their engagement with him at the expiration of one year from that time.

With the turn of the year Colonel Fulford gave notice at the Epiphany Sessions 1841 of a motion to be presented at Easter for the appointment of a general bridge committee. At last the sequence of ad hoc committees appointed as necessary for major events such as renewing the five year contracts for repairs was to be succeeded by a general bridge committee to receive and investigate the report of the County Surveyor prior to each meeting of the Quarter Sessions.

4.22 BRIDGES RENEWED AND REPAIRED 1831–1841

During the decade five three-span bridges were built, Bellever Road, Polson, Long Bridge, Plympton, Crocombe and Newton Poppleford. All were over rivers that were prone to sharp and substantial floods, the Dart, Tamar, Plym, Teign and Otter and all have withstood the elements for more than 160 years with minimum maintenance being required. Some of the larger new single-span bridges were Creedy, Cadover, Templeton, one over the river Coly at Colyford and one over the river Axe at Colyford.

Apart from Cadover, all the arches have a distinctive segmental design with a rise of only one sixth or one eighth of the span. This is a large ratio for masonry bridge construction and it is a tribute to the soundness and solidity of the abutments that the arches have not spread or settled in any way.

Such a design gave a maximum waterway free from obstruction for a particular span. Creedy Bridge, with a rise of 3 feet 6 inches for 25 feet (1 in 7) was as daring as any. Green supported his abutments with wing walls parallel to the road and a buttress midway along the back of each abutment.[13]

Nevertheless, Cadover Bridge, a single span over the river Plym, was quite different. Its arch sprang from close to the river and the rise of the arch was much greater so that the segment was close to being a semi-circle with the crown of the arch well above flood level. Another difference was in the siting of the wing walls. Hitherto Green had always made his wing walls align with the face of the arch in order to extend the parapet on each side with a slight outward curve at each end as though to gather the highway traffic to make the crossing. At Cadover he aligned his wing walls with the river but set them behind the pilasters so that they were well behind the river bank.

Unusually, Green designed a timber bridge over the river Otter at Tipton St John in 1833. This has masonry abutments and spanned 50 feet in two spans supported on a central trestle of timber. There were three longitudinal timbers 14 inches deep by 10 inches wide and 3 inch planking for the roadway which was 10 feet wide and had outriggers to brace the eight sets of handrail supports. The abutments, 3 feet 6 inches thick, were taken to a depth of 5 feet below river bed level. The contract was due for completion on 1 September 1834.[14]

Three spans, the centre span of Polson Bridge, the single spans of Iron Bridge, Landcross, and Axe

Bridges, Canals and Newport Dock

Axe Bridge, Colyford. This modern aerial view shows the difficulties this ancient bridging place presented.
(SY259926)

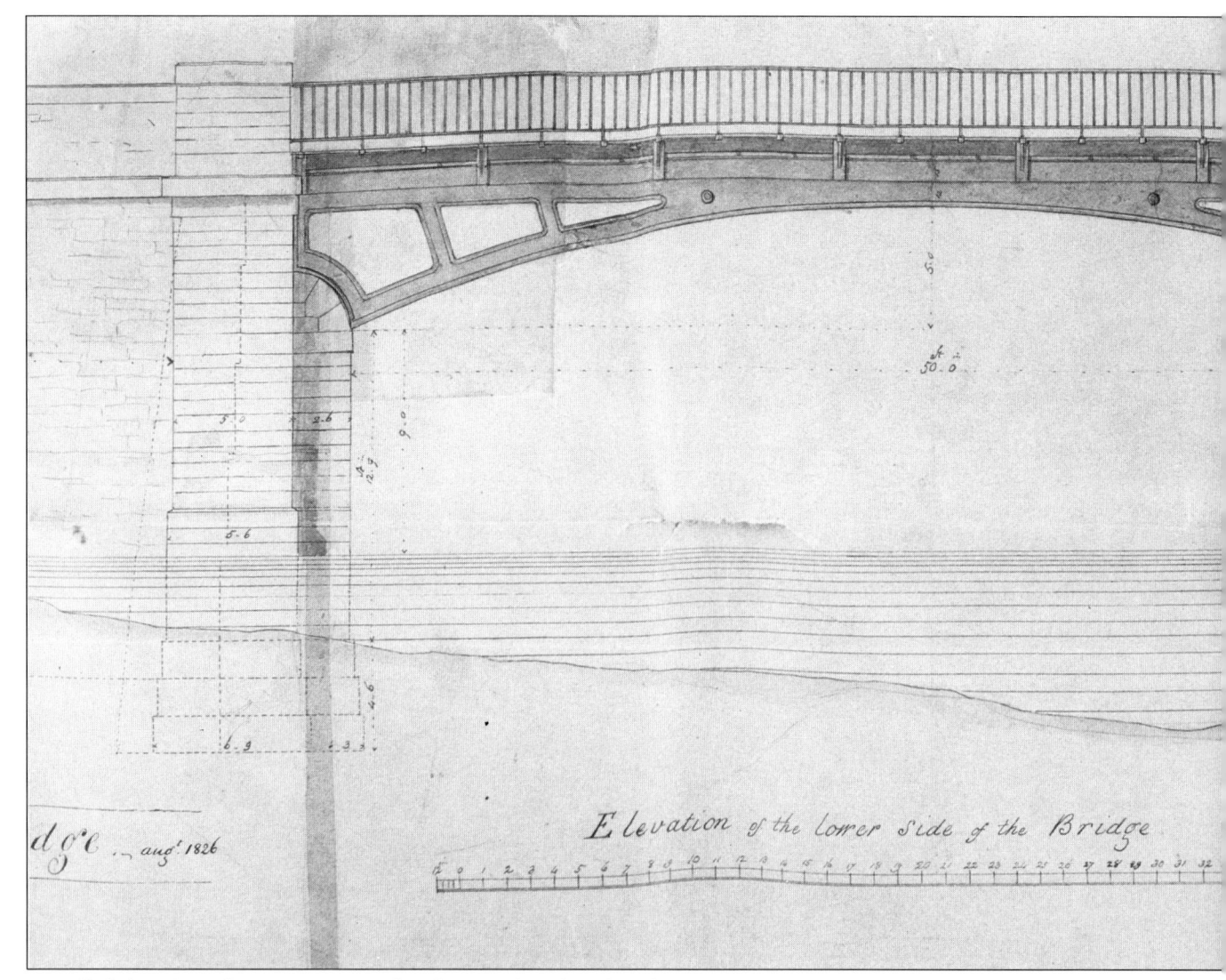

Detailed drawing of cast iron span, dated 1826, but found with Axe Bridge, Colyford contract documents and conforming with Whitaker's drawing

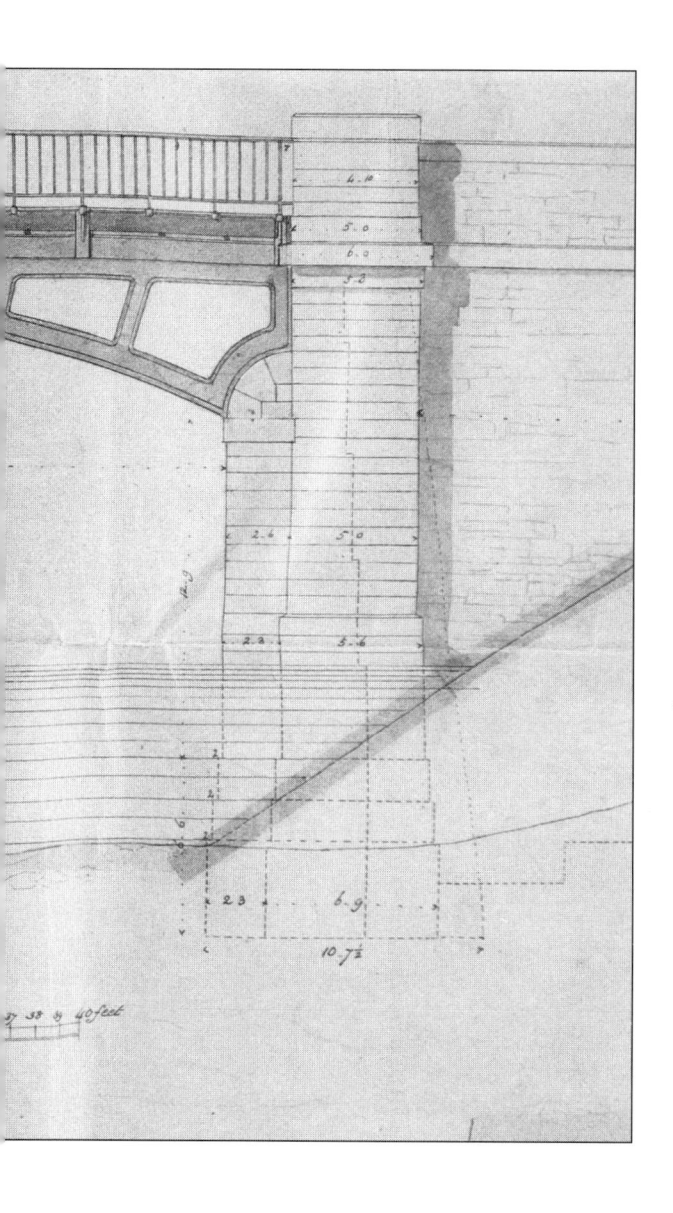

Bridge, Colyford, were of cast iron construction. Green required a span of 50 feet and at Polson he used six cast iron ribs at 5 feet centres to support the deck. The ribs were 6 feet deep over the supports and 12 inches deep at the centre with a segmental curve to the bottom portion of the rib while top and bottom portions were separated by two struts in the deeper part to give an open appearance to the web. The iron was shaped at the ends to bed on a curved surface, presumably to allow thrust and support to be developed together.

Axe Bridge, Colyford, differed from Polson in having five ribs instead of six. The contract for the cast iron was signed by George Brewer of Newport, Ironmaster, on behalf of the Coalbrooke Vale Iron Company for delivery by 31 May 1837. Again the span was 50 feet with the segmental curve to the underside of the ribs rising 5 feet to the centre leaving the rib only 12 inches deep at the centre of the span.

Once again the ends of the ribs were shaped to bed on a quadrant surface 2 feet 6 inches radius and the open panels must have been a source of weakness. The roadway was 19 feet wide between pilasters with an iron handrail over the deck. The abutments, 24 feet wide, were of narrow vertical section varying from 4 feet 9 inches thickness to 6 feet 1 inch at the base with a return at the ends to carry the pilasters and a counterfort 3 feet 6 inches outstanding at the rear at the centre.

The cast iron designs have not survived, Axe Bridge being replaced about 1912, Iron Bridge, Landcross, in 1926 and the centre span of Polson in the 1930s.

The reason for this was partly the design and partly the excessive weight of vehicles since the turn of the century, the early steam traction engines having heavy axle weights. As regards design, the cast iron ribs were very slender in elevation, neither deep enough at each end to act as arches, nor deep enough in the centre to act as beams, and the lack of material in the webs meant that they had insufficient solidity to accept thrust and shear in the bearing area. Moreover 50 feet span was a length where temperature stresses in the cast iron could have been important.

For all that, a life of 75, 95 and 100 years for a cast iron bridge with rapidly increasing loads was a reasonable return on the investment. Axe Bridge, Colyford, had cost £750 for the ironwork and £1100 for the masonry. Few people in the early nineteenth century foresaw the increase in the weight of vehicles and whereas masonry arches started by having to sustain their own enormous weight, the new material required very careful design to cope with 150 years of increasing loads if they were not to depart from arching action.

Green completed 33 years of bridge building for the county, producing masonry arch bridges to a high standard of excellence in design and appearance that remains evident today. The only criticisms of his engineering would seem to be that he did not appreciate the tremendous forces developed behind high retaining walls, and in his use of cast iron for bridge decks he only designed for the traffic loads of the day. The latter would not be a fault in these days of designing to the British Standard loads, where over-design is considered a fault.

The quarter by quarter progression of construction and maintenance began at Epiphany 1831 with an order to build a bridge over the mill leat at Otterton for £75 and to build Shuttern Bridge with a contribution of £100 to be spent if the local inhabitants of the neighbourhood agreed to raise a subscription for a larger bridge than the existing one.

At Easter the court consented to the alteration of Bow Bridge, Ipplepen, without expense to the county but to the approval of the County Surveyor. However, at Michaelmas 1832, Green was reporting that the turnpike trustees had taken down the existing bridge without notice to him or application to the Sessions and were then rebuilding it. There was no further report so presumably that storm in a tea-cup was satisfactorily resolved. Approval was given to the rebuilding of Bellaford (Bellever) Bridge for £120 provided that it would be at least 9 feet wide between parapets and that the parish would pay for any extra width.

At Midsummer 1831 the court ordered that £200 be paid towards the rebuilding of Wilmington Bridge provided that the remainder of the money was provided by the Axminster Turnpike Trust. A committee of magistrates were meeting a committee from Cornwall regarding the rebuilding of Polson Bridge over the Tamar near Launceston. At Bellever Bridge the ancient bridge was 'suffered' to remain and the contractor compensated for the stone that would have been available in the sum of £20. Two years later the treasurer was ordered to pay the Rev. James Holman, a mason, £120 for rebuilding the bridge and the £20 for the materials that could have been used from the other bridge. It

Bellever Bridge 1831 (SX658773)

seems, therefore that this lovely bridge was built for the cost of the masonry and that the mason must have provided his services free.

The County Solicitor reported at Michaelmas 1831 that Templeton Bridge was a county bridge. It was also in a dangerous state and the court ordered repairs at a cost not to exceed £70. The committee of the two counties considering Polson Bridge recommended a new bridge and Mr Arundel, a magistrate, received the thanks of the court for offering stone free of charge from his quarry. It was ordered that an advertisement should be made for plans and an estimate for the reconstruction of the bridge under the direction of a committee of six magistrates from each county.

Substantial sums were paid on the accounts of Wilmington and Winters Bridges at the Epiphany Sessions and the County Surveyor reported the erection of Eggesford Bridge in a substantial manner (which he had built for the Hon. Newton Fellowes in 1824) and this bridge was taken over by the county. Perhaps it was significant that Fellowes was chairman of the six Devon magistrates who were looking for a suitable person to build Polson Bridge. At the same Sessions the Teignbridge magistrates were directed to seek plans and specifications for the building of Jews Bridge.

At Easter 1832 Jews Bridge was ordered to be rebuilt at a cost not exceeding £850 and the Teignbridge magistrates were directed to enter into a contract, tenders having been submitted. At the same time a petition was accepted for the erection of a stone bridge over the river Erme at Caton. At Midsummer both Wilmington and Winters Bridges were certified as completed by magistrates of their divisions and Mr Green was directed to state his opinion of the plans that had been prepared for Polson Bridge.

Presumably this opinion contained reservations for at Epiphany 1833 the Hon. Newton Fellowes, on

Wilmington Bridge 1832 (SY218999)

behalf of the Polson Bridge committee, was reporting that the work would be carried out agreeably to the plans, specifications and agreement that day produced by Mr Green and signed by the committee of the two counties through their chairmen. The solicitors of the two counties were to prepare a contract 'agreeably thereto'. Mr Green had named William Crockett Esq. and John Harris Esq. of Exeter for the due performance of the contract, at the joint expense of the two counties, so here was Green acting as designer and contractor once again. The minutes noted that reasonable compensation be made to Messrs Hawkin for loss of time and attendance upon the committee so they may have been responsible for the plan upon which Mr Green had been asked to give his opinion!

At Easter 1833 Thomas Tucker was paid the balance of £139 out of a total of £989 having completed the

BRIDGES, CANALS AND NEWPORT DOCK

building of Jews Bridge over the river Bovey to carry the Exeter to Plymouth road. At Michaelmas the magistrates of the local division were ordered to meet the corporation of the Borough of South Molton concerning the repair of Bray Bridge. A new bridge was ordered to be rebuilt as soon as possible at Alphington for £250 and Loxbrook Bridge near Broadclyst for £120.

Meanwhile, in Midsummer 1833, £700 had been voted towards the costs of the Plymouth Turnpike Trust for the erection of a new bridge to replace

A painting of the Old Long Bridge, Plympton, prior 1835 (SX519567)

Long Bridge over the river Plym at Plympton, provided the County Surveyor and magistrates were satisfied as to the standard of its construction. However the Plymouth Eastern Turnpike trustees had second thoughts and at Epiphany 1834 presented a memorial suggesting widening the bridge at a cost of £574. The magistrate for the division, Mr Langmead, moved that in this case only £515 be requested from the county but at this proposal Colonel Fulford moved that the sum of only £300 be contributed by the court, and that again the work should be supervised by the County Surveyor.

Mr Green was ordered to report at the next Sessions on the practicability of the scheme proposed by Mr Langmead and at Easter he said that while it was practical to widen the existing Long Bridge he did not recommend it and moreover he did not approve of the plan submitted by Mr Langmead at the last Sessions. So Mr Langmead withdrew his motion to rescind the original Midsummer order.

Finally at Michaelmas 1834, after a year's delay, it was ordered that the court consented to the erection of a new bridge of limestone provided the trustees of the turnpike erected the bridge to the satisfaction of the magistrates and the County Surveyor and the sum of £500 was substituted for the £700 voted at Midsummer 1833. The County Surveyor's design prevailed and he kept his grip on the style of construction. At Epiphany 1836 work was reported to be well in hand and half of the £500 ordered to be paid to the trustees. It was reported completed at Michaelmas 1836.

Weatherwise, 1833 must have been a good year, for at Epiphany 1834 James Green reported that Polson Bridge was now completed and the County Surveyor of Cornwall, Mr Chapple, who had been appointed clerk of works by the joint committee, had presented a bill for £20 16s. It was ordered by the court that £10 8s be paid by the county of Devon.

At Easter 1834 the county Solicitor reported that he considered Farrants Bridge on the Moretonhampstead road to be the county's liability as a carriage bridge. The magistrates reported that they had two plans laid before them for rebuilding Tipton Bridge, one an iron bridge at an expense of £1400 and the other a wooden bridge for £400. The court ordered that Tipton Bridge be rebuilt in wood for £400.

Tailwater Bridge had been ordered to be repaired at Easter 1830 for £40. Mr Kennaway moved that £60 be added to this sum and the bridge rebuilt provided £50 could be raised by the inhabitants of the parish and neighbourhood, but this motion was lost on a division. At Midsummer 1834 a motion made by Mr Hull for widening Topsham Bridge agreeably to a plan presented by the County Surveyor at an expense of £800 was not seconded. This would have been the bridge over the river Clyst at Topsham, so presumably there were now two Topsham Bridges on the county list. The other bridge, over the river Avon and on the 1809 list, seems an unlikely candidate for this sum of money.

Templeton Bridge had been added to the list of bridges in 1831 with a vote of money for repair but at Midsummer 1835 it was ordered to be rebuilt at a cost of £300. At Michaelmas the County Surveyor was ordered to meet his Somerset colleague to consider the state of Exe Bridge, Morebath, and magistrates of the division to consult their opposite numbers in Somerset. The court refused to grant £50 to assist the Plymouth and Modbury Turnpike Trustees to repair Sequers Bridge but did agree to give them permission to widen and improve the approaches to the bridge provided it was done to the satisfaction of the County Surveyor.

In the year 1836 Green was busy providing new bridges. At Epiphany a new bridge was ordered for £750 at Crocombe over the river Teign and an attractive three-span bridge was provided. At Midsummer a new bridge was ordered over the river Axe near Colyford at a cost of £2000, a large sum of money for an iron span of 50 feet. At the same time Higher Creedy Bridge was to be rebuilt for £300 to carry the Crediton to Bickleigh road and Wrixhill Bridge between the parishes of Bratton Clovelly and Thrushelton for a cost of £280. The latter bridge was removed in the 1990s to allow the dual carriageway road to be built from Okehampton to Launceston.

Following the court's assumption of responsibility for Farrant's Bridge at Easter 1834, at Easter 1837 the trustees of the Exeter Turnpike gave notice that they desired to take down the bridge and rebuild it in the course of their improvement to the line of road. The court ordered that leave be given to the trustees to effect this improvement under the supervision of the County Surveyor and to the satisfaction of the magistrates of the division.

Crocombe Bridge 1836   *(SX848811)*

Higher Creedy Bridge 1836   *(SS846012)*

At Midsummer 1837 West Bridge, Tavistock, was ordered to be widened to a plan and specification of the County Surveyor and the trustees of

Plymouth Turnpike were called upon to pay £150 towards the expense. At Epiphany 1838 at Axe Bridge, Colyford, the magistrates of the division were invited to enquire whether £30 should be spent on the approaches. The difficulty of this site is that however commodiously the bridge is built, the highway approaches are little above Ordnance Datum and river floods combined with high tides still cause flooding of the approach highways. A similar problem was revealed in the minute noting that Topsham Bridge should be referred to the magistrates of the division in which this bridge was situated to ascertain the liability of the county to the repair of the bank and footpath within the county bounds of that bridge. If there should be any difficulty they were authorised to call in the County Surveyor for his advice on the subject.

At Midsummer 1838 it was at last ordered that Talewater Bridge should be rebuilt at a cost of £160 provided half the cost was contributed by the local inhabitants. The bridge had been considered in 1830 and 1834 but it was not until Easter 1839 that agreement was finally achieved that the £160 should be spent if the local inhabitants first contributed £60. First Bridge, Cullompton, had been washed away by the flooding of the river Culm and at Epiphany 1839 it was ordered that an iron bridge should be erected on this site at a cost not exceeding £600. It is now one of the few remaining cast iron structures erected by James Green.

Another long-standing need was acknowledged when, at Easter 1839, the reconstruction of the Newton Poppleford Bridge was ordered at a cost of

First Bridge, Cullompton 1839   (ST023067)

£2500. At last a bridge of sufficient size would be built across the river Otter to carry the Exeter to Lyme Regis road. Northbrook Bridge between Exeter and Countess Wear had been considered by a committee of magistrates and reported as being very inconvenient, in need of repair and the raising of the approaches. At Epiphany the committee was given power to erect a new bridge provided that the expense was no more than £200, that the trustees of the Exeter Turnpike agreed to this work and that no more than the £200 would be asked from the county.

At Easter the magistrates were authorised to employ a clerk of works to superintend the rebuilding of such parts of Newton Poppleford Bridge as would be underwater. A sum of £50 was allowed for this. At Midsummer 1840, the Sessions when James Green was given a year's notice of the termination

Newton Poppleford Bridge 1840–41   (SY091898)

of his appointment, it was resolved that it was inexpedient to repair Cadworthy (Cadover) Bridge over the river Plym and that £300 should be spent in rebuilding the existing site. Laverton Bridge was also to be rebuilt at a cost not exceeding £100 to the plan and specification of the County Surveyor.

The rebuilding of Newton Poppleford Bridge had run into trouble and the County Solicitor was directed to sue the sureties of the contractor immediately after 1 August unless terms could be made to satisfy the committee. The committee was also empowered to advertise for the completion of the bridge. At Michaelmas 1840 the County Surveyor reported that the tenders secondly procured all exceeded the sum granted by the Sessions. The matter was referred back to the committee for this bridge to consider what other plan could be used to effect completion for £2500.

Also at Michaelmas the County Surveyor reported that the contractor for Cadworthy Bridge required an extension to 21 December for the completion of his contract. This was agreed on condition that care must be taken that other parts of the contract were not prejudiced.

At Epiphany 1841 it was reported that a new contractor had been selected to complete Newton Poppleford Bridge and it was agreed that £10 should be paid to Mr William Hare for safeguarding the materials while work had been suspended. At the same time it was agreed that Colyford Bridge should be rebuilt at a cost not exceeding £500.

At Midsummer, as Green left the county employment, it was ordered that the weir should be repaired at Fenny Bridges for £150, that small additional sums should be paid to the contractor for Colyford and for Northbrook Bridges, and £150 be paid for extra work at Cadworthy (Cadover) Bridge. Cadworthy and Newton Poppleford Bridges were completed under the supervision of Mr Whitaker.

4.23 THE COUNTY BUILDINGS

During this decade there was no change in the county buildings in Exeter centred on the castle within the city walls, the gaol together with the house of correction and the sheriff's ward.

The County Surveyor presented a specification for the repair of buildings to the Epiphany 1831 Sessions and the visiting magistrates were instructed to examine this with the assistance of the county officers and, when the specification was

Cadworthy (Cadover) Bridge 1840–41   *(SX555646)*

approved, to advertise for tenders for the work which would be submitted to the next Sessions.

The dangers of lead piping must have been apparent because at Michaelmas 1831 it was asked whether iron could replace the lead pipes at the house of correction. The County Surveyor was ordered to assist a committee considering ways of improving the benches in the castle. A motion was put forward by Mr Sillifant and taken over by Mr Fellowes to give leave to the Exeter Turnpike Trust to form a road through the county fields in front of the county prisons (to eventually be called New North Road).

At Midsummer 1832 the visiting justices noted that juvenile offenders were of necessity being employed on the treading wheel with the older offenders, and the County Surveyor was directed to prepare a plan and estimate for an addition to the treading wheel that would enable the juveniles to be separated from the older prisoners.

In 1833 the County Solicitor reported his valuation of the Gaol Fields land for the new highway at £225 while the magistrates contracted with Charles Stokes of Heavitree, mason, to build a bridge across the brook at the bottom of the Gaol Fields for £24 9s 10d. The County Surveyor was instructed to advise and prepare plans and specification for altering the chapel at the gaol. At Michaelmas Baldwin Fulford jnr., chairman of the visiting justices, presented 'a plan for uniting the two chapels of the Gaol and Bridewell into one chapel at the Gaol capable of containing the whole of the prisoners upon three floors ... the plan appears to present the advantage of complete classification (of the prisoners) and at the same time to place all the prisoners under the view of the chaplain and all the officers – it would also enable the prisoners to have all the advantages of two services on each Sunday – the expense is estimated at £460.' The order to proceed was given at the Epiphany Sessions of 1834.

In 1834 the bridge at the bottom of the Gaol Fields was completed after the magistrates had made some alterations and Mr Stokes was paid £17 2s 2d. At the same time the visiting justices drew attention to the drain at the bottom of the fields (i.e. the Longbrook) and entered into talks with the Exeter Improvement Commissioners, no doubt with the cholera outbreak in mind. It was considered necessary to cover the brook through the Gaol Fields.

The Exeter Improvement Commissioners wanted to cover a 570 feet length at a cost of £423 but the visiting magistrates noted that the length of the drain within the county boundary was 345 feet, the remainder being the property of the Chamber. The justices estimated a cost of £237 10s which would largely be reimbursed by the sale of the land for New North Road to the turnpike trustees!

As for the juveniles under 15 working on the treadmill, the justices decided to take them off this punishment and communicate with the surveyor of the turnpike trustees to provide a quantity of stone at the Bridewell for the juveniles to crack for the use of road construction and maintenance.

The County Surveyor was instructed to examine the drains in the prison and their connections to the drain in the fields and by Easter 1833 Messrs Hooper had been paid £37 4s for altering the drains at the prison and covering a portion of the drain at the bottom of the fields. By Michaelmas a contract had been let to Mr Clarke at £169 for the main work to the drain in the fields and work was in progress.

Two other important matters were reported at the Michaelmas Sessions 1833. A proposal for a county lunatic asylum had been made at the previous Sessions and now the committee estimated a cost of £15,000 for building and land for 250 patients to be provided under the Act of Parliament 9 Geo IV. This would eventually be built at Exminster in the 1840s. By another Act of Parliament the prison governor would no longer be appointed by the Quarter Sessions court but by the Secretary of State and all rules in force would be annually reported to the Home Office. Power had been given to the Secretary of State to appoint not more than five inspectors for all the prisons throughout the United Kingdom. A House of Lords report had recommended one rule of law for all prisons and that separation and silence should be strictly enforced.

Building work now began to tail off. At Epiphany 1837 a chimney stack in the sheriff's ward was reported to be unsafe but the County Surveyor could find no distress and estimated that lowering it would cost £5. Nevertheless it was ordered that a particular examination should be made by the County Surveyor of all the ward chimneys shaken by a recent storm and if necessary the visiting justices would have power to carry out any necessary alterations.

Two years later, the report of the Inspector of Prisons, stating that a wall should be erected to divide the male and female prisoners in the sheriff's ward, caused the County Surveyor to lay before the court a plan with an estimate of £35 and this proposal was adopted.

At a meeting of the Exeter Turnpike Trust on 5 December 1838 it was reported that a letter had been received from Mr Green regarding the fences of the county fields adjacent to New North Road. In January 1839 the trustees agreed to put into repair these fences and at the Easter Sessions the visiting justices suggested that the court should order the work to be carried out without delay and not accept an offer of £62 compensation. On 1 April 1840 a letter from Mr Green to the trustees called for further repair to the Gaol fences presumably because this work had not been entirely adequate. The trustees accepted a tender of £6 6s and they ordered the work to be done.

4.24 THE BARNSTAPLE BRIDGE TRUST

The history of Barnstaple Bridge can be traced back to c.1280 and to 1333 when, during its repair with masonry, workmen were drowned.[15] The 16-span bridge was first widened in 1796 by the construction of outer segmental arches, springing from the cutwaters when the width was increased by 2 to 3 feet on both upstream and downstream sides.

By the 1830s pedestrian and horsedrawn traffic must have been considerable for the 12 feet width between parapets and the bridge trust (it was not a county bridge) engaged James Green to further increase the capacity of the bridge. This was done by cantilevering ironwork supports for a 4 feet wide footway with iron handrails on each side of the bridge, leaving a 16 feet wide roadway between kerbs. This took place in 1832–34, the ironwork being supplied by the Neath Abbey Iron Company and cost £4000. Henry Hebberley Price, a partner in the Neath Abbey Ironworks, had also been brought up in the Quaker tradition and was recorded in 1838 as a member of the Institution of Civil Engineers. A rectangular plaque of cast iron on the corner of Bridge Chambers adjacent to the north-east end of the bridge has the following inscription:

> This bridge was widened with ironwork by the Neath Abbey Iron Company in the year 1834, James Green, Engineer.

These iron cantilevered footways lasted successfully for 130 years until ownership of the bridge was transferred to the Ministry of Transport and the bridge widened again in the 1960s. The delicate appearance of the handrails enhanced the appearance of the bridge even if they gave less protection to pedestrians from the wind. Many local people were sorry to see them replaced by a masonry parapet again. However, the 16 feet wide road and the relatively fragile footway gave no protection from huge lorries should one mount the footway.

4.3 WATER SUPPLY, SEWERAGE AND RAILWAY PROPOSALS FOR TORQUAY

Green proposed a scheme for the water supply and sewerage of Torquay in 1832. A dam and reservoir were to be situated on the Fleet brook just above Castle Circus. From an intake two parallel mains, one for water supply and one for a sewer, led down the line of the present Union Street and Fleet Street. The water main had branches to Vaughan Parade and Torwood but otherwise led along the Strand into Victoria Parade. The sewer followed the water main along the Strand into Victoria Parade to discharge into the outer third of the old harbour of which the inner two-thirds was to be protected by a harbour arm and a lock gate. The deposited plan was to a scale of 2 chains to 1 inch.

At the same time he proposed a scheme for a railroad from Newton Abbot to Torquay. Leaving Newton Abbot in the area of East Street it curved north and east around the contours towards Decoy Road to pick up the route of the Aller Brook via Langaford Bridge, Aller Bridge, Kingskerswell, Scots Bridge and Tor to the harbour where it turned into Vaughan Parade. There were sharp horizontal curves in Newton Abbot, one rising gradient of 1 in 36 and two descending planes in Torquay of 1 in 9 separated by a tunnel. No indication was given of the method of powering the railroad vehicles, but the 1 in 9 planes suggest stationary steam engines with horsepower elsewhere, rather like his canals of the previous decade in North Devon. The deposited plan and longitudinal section were to a scale of 10 chains to 1 inch.

Both plans and section were signed by James Green and countersigned by Wm Kitson and the Clerk of the Peace, J.E. Drake, on 30 November 1832.[16]

4.4 CANALS AND DOCKS

4.41 GRAND WESTERN CANAL

As mentioned in chapter 3, the idea of a link between the Bristol and English channels had been alive since 1768 and a particular route from Taunton to the river Exe at Topsham via Burlescombe had been proposed by John Rennie in 1794. This had been resisted by the mayor and chamber of Exeter because their recent expenditure on the Exe canal was far from being recouped and, also, the use of river Culm water for Rennie's proposal would have reduced the water supply to the river Exe. This and the effect of the war with France caused the proposal to be abandoned until interest in the Grand Western Canal proposal was revived in 1810 with Rennie as engineer.

This time the need for a canal along two and a half miles of the main route and nine miles of the projected branch to Tiverton was recognised so that limestone could be conveyed from the Canonsleigh quarries to Tiverton, where kilns would be installed. Water supply was provided by springs at Lowdwells and the summit was lowered by 16 feet so that a level canal could be provided suitable for barge traffic. The canal was opened in August 1814 at a cost of £250,000.

During the 1820s, Green was involved in other proposals for canals between the Bristol and English channels but eventually the need for cheaper coal to burn limestone at Tiverton led to the Grand Western canal proprietors wishing to extend their canal to Taunton to join the Bridgwater and Taunton canal. The intervening distance was not great at 13 miles but the difference in levels of 262 feet was an obstacle as it would appear to require a multitude of locks and a large supply of water for their operation. John Rennie had died in 1821 but his former assistant James Green had become expert in the construction of canals in hilly country with the building of the Bude Canal using inclined planes and tub-boats of 5 tons with wheels attached to them to run on the inclined planes.

At a meeting on 1 May 1829 Green presented a report to the Grand Western Canal Company observing that while the Lowdwells to Tiverton canal was 6 feet 6 inches deep, 24 feet wide at the bottom and 46 feet wide at the top with a clear waterway under the bridges of 18 feet, in the Bridgwater to Taunton canal the width under bridges was only 14 feet and so the width of any new canal need be no more than this. Green recommended 13 feet width at the bottom, 23 feet at the water surface and 3 feet depth. He advocated boats 20 feet by 6 feet, carrying 5 tons and drawing 2 feet of water. Six of these could be drawn by one horse on the level and could navigate the Bridgwater and Taunton canal in sets of six. Using inclined planes instead of locks, his estimate for the cost of the new canal was £50,000, a great reduction on the cost of the 1814 length.

In a further report of March 1830 Green made some changes suggesting there should now be one inclined plane and seven perpendicular lifts and that boats of 8 tons capacity should be used. The estimated cost was advanced to £61,324. The change to lifts was due to the very gradual rise of the land over the first five miles from Taunton and the need to restrict the use of water.

Locks, lifts and planes constructed were as follows:-

Taunton	stop lock	
Taunton	lift	rise 23$^1/_2$ feet
Norton	lift	rise 12$^1/_2$ feet
Allerford	lift	rise 19 feet
Trefusis	lift	rise 38$^1/_2$ feet
Nynehead	lift	rise 24 feet
Winsbeer	lift	rise 18 feet
Wellisford	inclined plane	rise 81 feet
Greenham	lift	rise 42 feet
Lowdwells	lock	rise 3 feet
	Total rise	262 feet

For his aqueduct over the river Tone at Nynehead, Green used a cast iron trough and for his highway

River Tone aqueduct (photo B.J.Murless)

bridges at Trefusis and Silk Mills he used light cast iron beams. Trefusis Bridge is of similar design to the Silk Mills Bridge. A record of the construction of this bridge has been made by D.J. Greenfield for the Somerset Industrial Archaeological Society.[17]

An advertisement was placed for canal contractors, masons and builders to respond to the invitation to tender for eight lots of work in December 1830. Eventually lots 1 to 5 were let to Messrs Houghton and Company and lots 6, 7 and 8 to H. McIntosh.

Work was reported as having commenced in June 1831 and by the middle of 1832 was reported by Green as well in hand. However in 1833 progress was lagging behind, particularly the three miles from Taunton, and the report for 1834 was also disappointing. The committee report of 1835 regretted that the true cause of the delay appeared to rest in the novelty of the lifts where many practical difficulties were only gradually becoming apparent.

Green had been engaged to complete the lifts for a certain sum of money and the many alterations necessary had been made at his expense. The canal was then open as far as Wellington, but by 27 January 1836 Green had ceased to be engineer to the company.

The particular practical difficulties with the lifts are well described by Helen Harris in her study of the Grand Western Canal where she also relates the report of the engineer, W.A. Provis, who was engaged by the canal company in May 1836 to investigate the lifts and inclined plane and the cause of the plane's failure. This he did by 28 June. Provis reported on all the engineering works and generally found the bridges, culverts and aqueducts to be in good condition, even commenting that the essential work was excellent.

However, Provis noted bulging of retaining walls in certain places and rapid loss of water in the canal below Allerton lift where the water was very low because there was no clay lining in the gravelly deep cut.

As regards the lifts, the first three from Taunton suffered from the water level in the caisson chamber not being able to be drained away properly and therefore remaining above the level of the lower canal pond. This meant that locks had had to be constructed at the lower end of the caisson chambers, thereby slowing the passage of boats. There had been some mechanical problems due to the breaking of shafts or pinions. He noted that the time taken to pass through the $12^{1}/_{2}$ feet Norton lift from lower to upper levels was $8^{1}/_{2}$ minutes.

The largest problem had been the performance of the Wellisford incline. Here an ascent of 81 feet in a distance of 440 feet was achieved over a gradient of 1 in $5^{1}/_{2}$. Here the incline differed from those at Bude in that two railway tracks had been laid so that boats could proceed upwards and downwards by being placed in a wheeled cradle to be carried up or down. At Bude wheels had been fixed to the boats but they had been the cause of damage to the sides of the canal, or had caught the sides and impeded the progress of the boat train along open cut.

The weight of a cradle and a boat was markedly different from those at Bude and experience with the bucket at Wellisford showed that the 8 ton buckets were too light as 4 tons of water was required to move the machinery without the cradles and 8 tons was necessary to overcome friction in the system merely to move empty cradles up and down the incline. If, therefore, there was no descending traffic and 8 tons of barge and contents had to be moved up the incline, there was

Bridges, Canals and Newport Dock

no reserve of power to achieve this and Provis calculated that 25-ton buckets would have been required. At Hobbacott on the Bude Canal 15 tons of water had been needed to raise boats weighing 6 tons. A 12 hp steam engine costing £800 was required and obtained to operate the Wellisford incline.

On 28 June 1838 the Grand Western Canal was at last fully opened at a cost of £80,000, some £30,000 above the original estimate, but at far less cost than the Lowdwells to Tiverton length. The canal remained in use between Taunton and Lowdwells until 1867 and from Lowdwells to Tiverton until 1924, but was never successful financially. Nevertheless the lifts operated until the closure of the canal in 1867 to the Devon/Somerset border, which was a remarkable achievement for a technology that had never before been successfully applied in Britain, and was not to be again until the Anderton lift was established between the river Weaver and the Trent and Mersey canal in 1875. Green published a paper on his canal lifts in what was only the second volume of the *Transactions of the Institution of Civil Engineers* in 1838, together with three drawings (see 4.8).

The failure of his design for the Wellisford incline was inexplicable bearing in mind his successful work with the Hobbacott incline on the Bude Canal. But his agreement to be financially responsible for problems with the lifts showed a man of character who was perhaps too ready to back his own proposals financially. Unfortunately this failure must have contributed to his bankruptcy in 1837.

In his paper, D.J. Greenfield described the measurements of Silk Mills Bridge, one of Green's road bridges over the canal, which was removed by the Somerset County Council in 1977 during road improvement works.

A cursory glance at the exterior of Silk Mills Bridge would have suggested that the road was carried over the canal on a very flat masonry arch. However the visible masonry arch supported only the parapets; arched cast iron ribs carried the roadway itself. Those five ribs spanning 3.9m and spaced 1.25m apart, were seated on a cast iron cill beam along the top of each abutment and had bolted on them cast iron road plates on which the

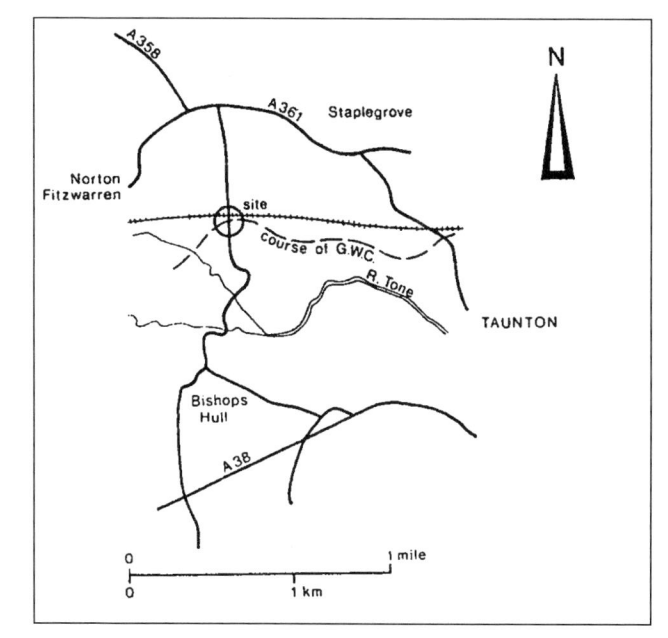

Silk Mills Bridge, location 1836

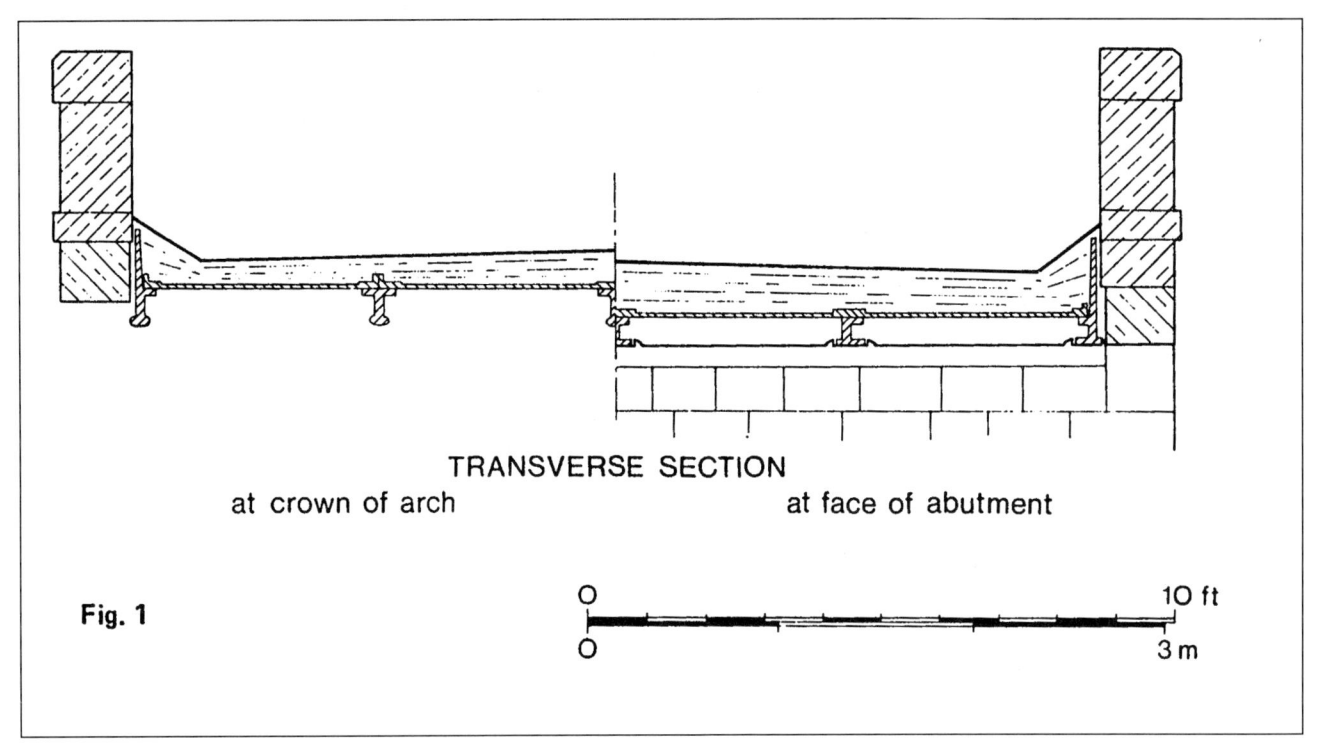

TRANSVERSE SECTION
at crown of arch   at face of abutment

Fig. 1

Silk Mills Bridge, transverse section 1836

actual road construction was formed. The two outer ribs had extended webs to support the road at the sides. The bridge was built to a skew of about 10 degrees, and the ends of the ribs and their seatings on the cill, together with the road plates, were cast to this skew.

The demolition of Silk Mills Bridge leaves extant only one cast iron road bridge over the line of the Grand Western Canal, that at Trefusis Farm, Bradford-on-Tone (ST166232). Two other structures containing cast iron survive which were once associated with the canal. These are the Nynehead Court Drive aqueduct (ST144218) and the river Tone aqueduct nearby (ST147224) which retain their cast iron troughs in which the canal was carried over a road and river respectively.

4.42 CHARD CANAL

Mr W. Hanning had suggested in 1831 that the Bridgwater and Taunton Canal Company should canalise the Westmoor main drain to provide a canal from the river Parrett to Ilminster and Chard.

Bridges, Canals and Newport Dock

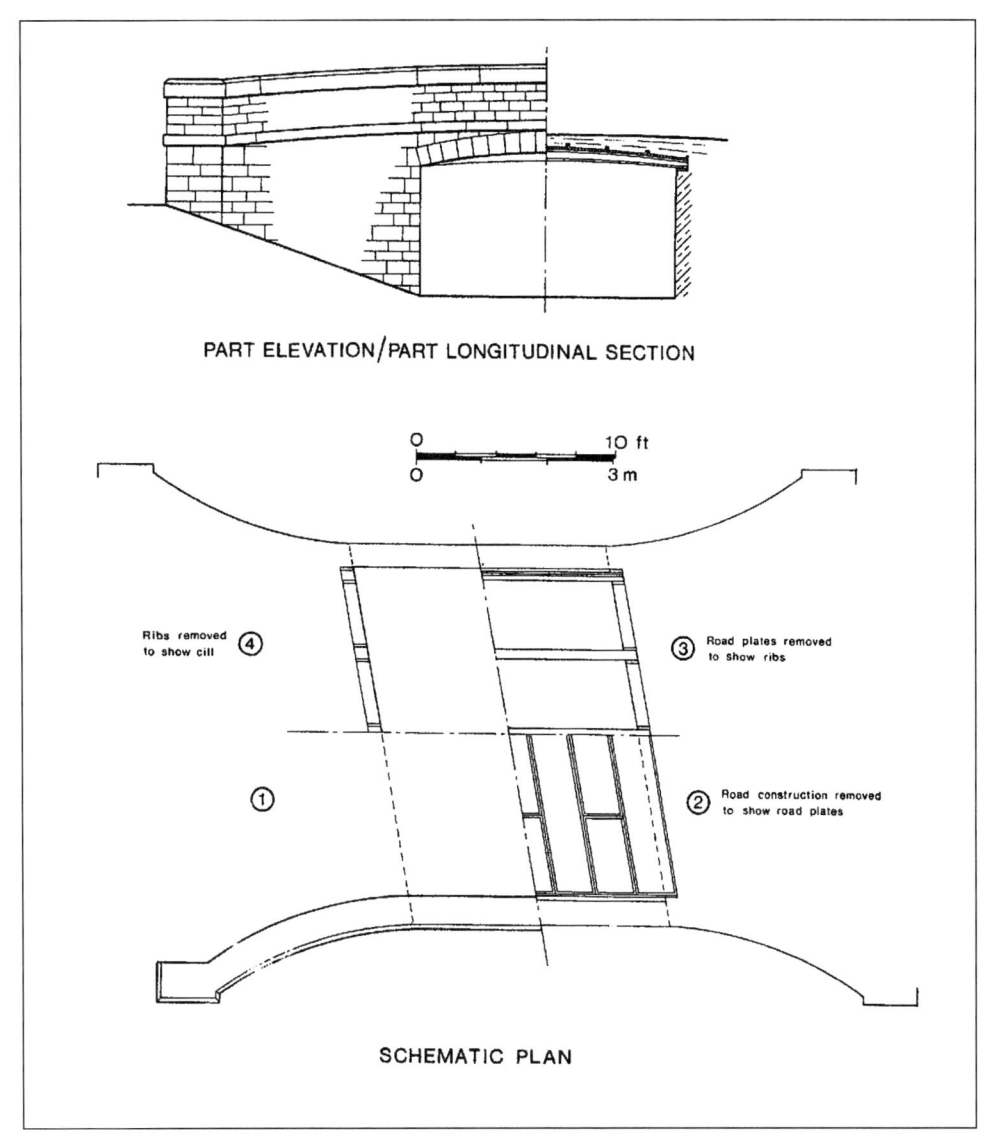

Silk Mills Bridge, elevation and plan 1836

James Green, who was engineer for the Westmoor drainage scheme, suggested that it would be more sensible to provide a canal direct from the Bridgwater and Taunton canal, to run from near Taunton to Chard. Green had obtained comprehensive information on a route from east of Taunton to Chard when he surveyed the line from Stolford to Beer for Thomas Telford, signing the plan for that scheme in 1824.

Now Green proposed a canal to leave the Bridgwater and Taunton canal at Creech St Michael. From this junction it would climb 231 feet over 13½ miles to reach Chard. Designed for tub-boats 26 feet long by 6½ feet wide, two lifts, two inclined planes and two tunnels would be required and the cost would be £57,000. Work got under way on 24 June 1835 but not throughout the route for, although an Act had been obtained in June 1834, land could not be bought compulsorily until all the £57,000 had been subscribed. Soon after commencement Green ceased to be engineer, no doubt because of the difficulties with the lifts and inclined planes on the Grand Western Canal.

The lifts were never built and the inclined planes moved boats in caissons rather than cradles so that the descending caisson could be full of water as counterbalance for an ascending boat in an empty caisson. This resolved the power problem at last. Presumably Green would have been reimbursed by the Bridgwater and Taunton Company for his design work for the Chard canal. The four-span aqueduct over the river Tone at Creech St Michael has all the appearance of the three-span bridges that Green was building for the county of Devon in the 1830s.

The canal was eventually opened to Ilminster in July 1841 and completed to Chard by May 1842, seven years after commencement.[18]

4.43 BURRY PORT AND THE KIDWELLY AND LLANELLY CANAL

The silting of the Gwendraeth estuary in Wales in the early nineteenth century had caused Kidwelly to lose its facility as a port for the coal of the Gwendraeth valley. Thomas Gaunt had developed a harbour at Pembrey and at one time Rennie had suggested that Gaunt could use the Kidwelly and Llanelly canal for access to it, but that company thought otherwise and by an Act of 10 June 1825 obtained powers to build a new harbour at Pembrey, later called Burry Port harbour. Pembrey New Harbour (Burry Port) was opened for traffic in 1832.

The Kidwelly and Llanelly Canal and Tramroad Company had obtained powers by an Act of 20 June 1812 to extend a canal up the Gwendraeth valley to beyond Cwm Mawr about 250ft above sea level. In 1832 the company called in James Green to report on extending the canal beyond the point reached in 1824 near Pontyates, five miles below Cwm Mawr and below the 50ft contour. No doubt this was because of Green's advocacy of the use of inclined planes.

Green reported on 29 July and on 29 October 1833. The first part of his report dealt with an examination of the existing work and the need for improvement, especially regarding its extension to the new harbour. The second part of the report recommended the extension of the canal from Pontyates to Cwm Mawr. He recommended two locks between Pontyates and Ponthenri and then three inclined planes at Ponthenri, Capel Ifan and Hirwaunissa at an estimated cost of £35,845.

By 1834 work was well advanced and £39,700 raised. However, in October 1835, Green was in trouble as he had had to inform the directors that he was not able to finish his inclined planes. He appears to have contracted for this work and the company had to continue the work and was forced to raise £8000 from the shareholders to complete.

Green was dismissed as engineer on 30 January 1836, three days after his dismissal from the Grand Western Canal Company. He was also dismissed a month later as engineer to Burry Port as a dock wall had collapsed and the failure of the walls caused the harbour to be closed for much of 1836. The Kidwelly and Llanelly Canal Company, already enraged by his failure to produce the inclined planes, could have had little sympathy with another failure and may have already been seeking to recover their additional costs from Green.[19, 20, 21, 22]

4.44 SURVEY FOR AN ALTERNATIVE LONDON TO BIRMINGHAM CANAL

One effect of the prospect of railway travel was that the canal companies looked for improvements to their own systems. In 1833 a group of promoters planned a new London and Birmingham canal and the line, which was surveyed by Green, was to run from the Stratford-upon-Avon canal to pass through Banbury, Buckingham, cross the Grand Junction near Tring and run to St Albans and Highgate, where a branch was to link up with the Regent's canal.

In all it was to be 113 miles long with 48 locks, and a further 22 locks on the Regent's canal branch. From the Stratford canal there were to be 19 locks and then a level pound of over 70 miles to St Albans. Nearly 10 miles of tunnelling would be needed as well as an enormous aqueduct over the Avon valley and the cost was estimated at £3,000,000. Where tunnelling was necessary, twin tunnels, each with a towing path, were to be made. Elsewhere the sides of the canal were to be walled and a double towing path provided throughout.

Comprehensive proposals were put forward in 1836 and there was a final meeting in Cubitt's London office in March 1838, but no more was heard thereafter.[23]

4.45 STOURBRIDGE CANAL EXTENSION

Early in 1836 the Stourbridge Canal Company resolved to ask James Green of Exeter to make a new survey from the summit of their canal to Gornal with a report as to the expediency of connecting it to the mining district. The survey was made, though apparently by William Fowler and not Green, and resulted in a proposal to link the Stourbridge to the Birmingham canal.[24]

4.5 LAND RECLAMATION, WESTMOOR, SOMERSET

Westmoor is about four miles from Ilminster towards Langport, lying between the rivers Parrett and Isle in the parish of Kingsbury Episcopi (ST4121). The 1811 Ordnance Survey map showed it as being marsh land surrounded by higher ground. There is a survey of Westmoor in the Somerset Record Office dated 1824 but unsigned.[25] The drainage of the southern Somerset Levels stood apart from the rest of the region in their late reclamation and in the abandonment of gravitational drainage and the adoption of steam pumping.

James Green was engineer to the Westmoor Enclosure.[26, 27] A Mr W. Hanning who supervised Green's work for the Ilminster Turnpike Trust and who had interest in the Westmoor area had given notice in 1831 of his intended canalisation of the Westmoor drain but this did not in fact come about and the Westport canal from the river Isle along the northern side of the Westmoor area was completed about August 1840.

Michael Williams has commented that the drainage of Westmoor in 1833 was probably the most advanced and comprehensive scheme of moorland reclamation ever attempted in the Levels.[28, 29] A catchwater drain was dug to intercept the run-off from the surrounding uplands and it had outfalls at either end which dealt merely with that run-off water. In the moor itself the usual network of rhynes was laid out and all rhynes led ultimately to a main centre rhyne $3^1/_2$ miles long, which ended in a clyse (tidal sluice) at the junction of the rivers Isle and Parrett. Irrigation was also considered and supply drains were connected to the catchwater drains from a weir at Slab Gate (ST416 234) on the river Isle and a weir at Combe Bridge (ST434 217) on the river Parrett. These could be regulated to release water throughout the moor in dry weather.

Green's work on the Westmoor drainage led to him being recommended as engineer for the Chard canal (see 4.42).

There is in existence a bridge drawing for a structure at Westmoor signed by Green on 17 November 1834.[30]

4.6 THE EXETER TURNPIKE TRUST

On 7 December 1831 the Exeter Turnpike Trust surveyor (Mr Wm McAdam) was authorised to see Mr Green and others who could give any information respecting the intended New North Entrance Road which was being designed to enter Exeter avoiding St David's Hill. This was probably because New North Road eventually skirted the grounds of Elmfield House on the eastern side and Green was well aware of the site, formerly his home.[31]

At the Easter Quarter Sessions in 1837 Green was asked to supervise the rebuilding of Farrants Bridge for the trustees as they intended to improve the line of this road. In March 1838 the trust ordered the sum of £50 to be paid to Mr Green for his plans and sections of an intended improvement at Farrants Hill and Warnscombe Hill on the road from Pocombe Bridge to Moretonhampsted (B3212). Nothing more than minor straightening of the road appears ever to have been carried out for there are still 1 in 7 inclines to the summit and thence down to Reedy.

Green's work had apparently been done at the request of the committee appointed for the improvement of that road without his estimated fees being notified at a general meeting. Col. Fulford had to request payment at an annual meeting before the £50 could be paid and the entry in the order book restated the rules about obtaining powers from a general meeting for such purposes. The fee was paid on 17 April.[32]

In December 1838 Green was writing as surveyor on behalf of the magistrates calling for repair of the fences of the county fields on the New North Road where the road passed the county gaol. On 1 April 1839 the trustees approved a tender for the necessary work of repair.[33]

4.7 RESIDENCES AND BANKRUPTCY

In 1830 and 1831 Green's home, as recorded in the *Exeter Pocket Journal*, was at 38 Southernhay Place while in 1833, 1834 and 1836 it was in Magdalen Street. In these years he was having to divide his time between the inspection and construction of the county bridges and attendance on the work of the Grand Western Canal, the Kidwelly and Llanelly Canal, Burry Port, the survey of the Birmingham Canal and the drainage of Westmoor. Following his dismissal from the canal and port posts, he suffered a severe financial and professional reverse, presumably because having contracted for some of the work he now had to meet claims for some of the losses. He must have spent a considerable time away from home finding a new line for the London to Birmingham Canal, even if he employed assistants for the detailed survey.

A notice of bankruptcy of James Green appeared in the *Exeter Flying Post* on 9 March 1837[34] following this entry in the *London Gazette* on Tuesday 7 March:

> Whereas a fiat in bankruptcy is awarded and issued forth against James Green, of the city of Exeter, Civil Engineer, Dealer and Chapman, and he being declared a bankrupt is hereby required to surrender himself to the Commissioners in the said Fiat named, or the major part of them, on the 29th day of March instant, and on the 18th April next, at twelve at noon on each day, at the Half Moon Inn, in the city of Exeter, and make a full discovery and disclosure of his estate and effects; when and where the creditors are to come prepared to prove their debts, and at the first sitting to choose assignees, and at the last sitting the said bankrupt is required to finish his examination, and the creditors are to assent or dissent from the allowance of his certificate. All persons indebted to the said bankrupt, or that have any of his effects, are not to pay or deliver the same but to whom the Commissioners shall appoint, but give notice to Messrs Gidley and Kingdon, Solicitors, Exeter, or to Messrs Burfoot, No. 2 King's Bench Walk, Innertemple, London.

A further entry in the *London Gazette* on Tuesday 25 April resulted from the two Exeter meetings:

> Whereas the Commissioners acting in the prosecution of a Fiat in Bankruptcy awarded and issued forth against James Green, of the city of Exeter, Civil Engineer, Dealer and Chapman, have certified to the Right Honourable the Lord High

Chancellor of Great Britain, and to the court of Review in Bankruptcy, that the said James Green hath in all things conformed himself according to the directions of the Acts of Parliament made and now in force concerning bankrupts; this is to give notice, that by virtue of an Act, passed in the sixth year of the reign of His late Majesty, King George the Fourth, intitled "An Act to amend the laws relating to bankrupts;" and also of an Act, passed in the first and second years of His present Majesty, intitled "An Act to establish a Court in Bankruptcy", the certificate of the said James Green will be allowed and confirmed by the court of Review, established by the said last-mentioned Act, unless cause be shown to the said court to the contrary on or before the 16th day of May 1837.

In the *Exeter Flying Post* in June 1838 appeared the following notice of a meeting of Commissioners:[35]

> The Commissioners in a Fiat of Bankruptcy bearing the 28 February 1837 awarded and issued forth against James Green of the city of Exeter, Civil Engineer, Dealer and Chapman, intend to meet on 5 day of July next at 1pm at Old London Inn in the city of Exeter to make a dividend of the Estate and Effects of the said bankrupt when and where the Creditors who have not already proved their debts are to come prepared to prove the same or they will be excluded the benefit of the said Dividend and all claims not then proved will be disallowed.[35]

By 1838, Green had moved to Alphington, no doubt to economise. There appears to be no record of his discharge from bankruptcy in the Public Record Office. Probably the sums involved in his failed contracts were so large that he had no opportunity to recover them from his professional fees during the closing twelve years of his life. This would have affected his status in the Religious Society of Friends who would have disowned him because of his bankruptcy.

4.8 THE INSTITUTION OF CIVIL ENGINEERS – GREEN'S FIRST PAPER

In 1836 the first volume of the *Transactions of the Institution of Civil Engineers* was published and in this volume we find a membership list. It contained office bearers, members, corresponding members and associates. The description of classes of member was as follows:

> Members shall be persons who are or have been engaged in the practice of Civil Engineering. Corresponding Members shall be persons of the same descriptions who reside without the limits of the threepenny post. Associates shall be those whose pursuits constitute branches of engineering but who are not engineers by profession.

There were listed 47 members, 93 corresponding members and 100 associates. James Green of Exeter was entry 37 on the list of corresponding members. Joseph Green of Exeter was entry 38 on the list of associates.

In 1838 the second volume of *Transactions* was published and contained a paper and drawings by James Green on canal lifts. Very little work had been done in England at that time on this subject so it was a paper of considerable importance. The paper was titled *Description of the perpendicular lifts for passing boats from one level of canal to another, as erected on the Grand Western Canal*.

Green began his paper by noting that the lifts had not been designed to supersede the use of locks on canals in all cases, but for a peculiar situation, in which a very considerable ascent was to be overcome in a short distance, where the supply of water was inadequate for the consumption of common locks, and the funds insufficient for the execution of the work on a scale adapted to such locks.

He continued that the trade expected on the Grand Western Canal was chiefly the carriage of coal, culm for burning lime, and limestone, all of which could be conveyed in a train of four, six or eight small boats linked together and drawn by one horse. The chief objective was economy in cost and therefore to move slowly and to convey a large quantity. It was therefore desirable that the ponds of the canal between lifts should be as long as possible and the lifts as few in number as possible in order to prevent hindrance by too often detaching boats in order to pass them singly over the lifts.

The boats were built to carry 8 tons each, being 26 feet long and 6^1/$_2$ feet wide, drawing when laden 2 feet 3 inches of water, so that a canal 3 feet in depth was sufficient for their use.

A lift consisted of two chambers, similar to the chambers of a common lock with a pier of common masonry built between them, each chamber being of sufficient length and width to enclose a wooden cradle of sufficient length and width to freely admit one of the boats to float within. The cradle was furnished with water-tight rising doors at each end to contain the boat afloat within when the cradle was lifted.

The side walls of the chambers and the pier within them were carried up from the foundations, sufficiently below the bottom level of the lower canal, to the top bank level of the higher pond of the canal. Water was prevented from flowing into the chambers of the lift from both the lower and upper ponds of the canal by lift-up gates or doors.

The floors of the chambers were sufficiently below the bottom of the canal to allow the coil or gathering of the balance chain underneath the cradles, to leave the crossbeams of the timber on which the cradles rested when at the lower level uninterrupted by water, and a drain was laid from each chamber to prevent the accumulation of water beyond the height of the bottom of the cradle.

The sides of the cradles were well secured by wrought iron knees on the inside and were riveted to wrought iron straps on the outside. The ends of the cradles had cast iron frames corresponding with their cross-section, properly bolted and riveted to the timbers, which preserved the uniformity of the figure, and rendered them perfectly stiff and firm.

On the top of the walls of the lift a framing of cast iron was erected, consisting of twelve upright hollow columns, 9 feet high and 12 inches in diameter, which were secured to the masonry by strong

Canal lift, side view

wrought iron holding-down bolts. The columns were braced together by lateral and transverse beams and this framing supported a longitudinal cast iron shaft, 22 feet in length and 10 inches diameter, with couplings. The bearings were

Canal lift, transverse section

turned and seated on brasses. On this shaft were fixed three cast iron wheels, 16 feet in diameter, for the purpose of carrying the wrought iron chains

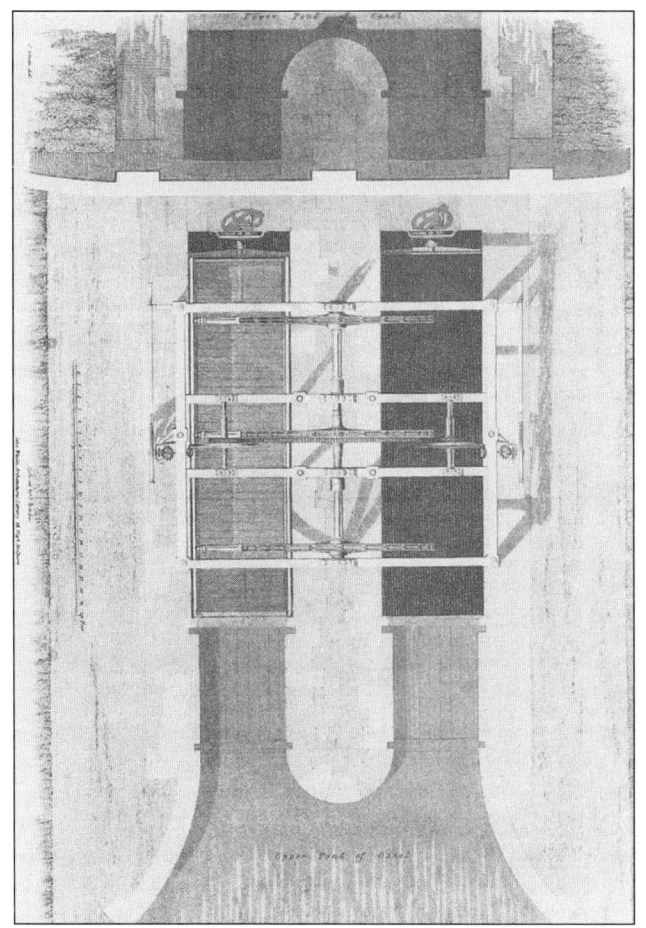

Canal lift, plan view

which supported the cradles, the extremity of the wheels being directly over the centre of each chamber of the lift.

The two outer wheels simply carried the chains, but the centre one besides this had a spur gear fixed on one side which worked into a pinion giving motion to bevel gear wheels which, by means of a wrought iron diagonal shaft, communicated with a hand gear or winch power fixed on the side of each chamber wall. The hand gear was attached to a brake wheel and brake lever for regulating the speed of ascent and descent of the cradles when a preponderance of weight of water in one cradle was used to assist motion.

The cradle was attached to the chains by means of strong wrought iron suspension bars fixed to each side in the plane of the three cast iron wheels. These bars were connected in pairs by a cast iron beam across the cradles, at a sufficient height above the sides to allow the boat to pass under them.

The length of the suspending chains was so arranged that when one cradle was at its proper level at the top, and each cradle containing an equal quantity of water, the weight would be in equilibrium but for the difference in length of the chain between the cradle at the bottom and the one at the top. To remove the disparity, chains of an equal weight per foot to that of the suspending chains were attached to the bottom of each cradle, with one end resting on the floor of the chamber, on which they gathered under a descending cradle, and elongated with an ascending one, so that an equilibrium was preserved.

Nothing more was wanted to put the machinery into motion than a power equal to its inertia and friction, together with the required velocity. This power was acquired by so adjusting the chains that when one cradle was at the bottom of the lift on the proper level to receive a boat, the cradle in the opposite chamber was not quite up to the level necessary to receive a boat from the upper pond of the canal. The difference found necessary in using the lifts was not quite two inches, producing in the cradle a preponderance of weight of only one ton, but this could be regulated as required. The cradles were so suspended that the facings at their higher end, when raised to the proper height, came within half an inch of the facings of the higher stop gate to the canal. The cradle was then forced forward close to the last mentioned facing by means of a forcing bar of cast iron at the rear of the cradle.

The advantages of the lift system over locks were economy in the expense of construction as compared with common locks, the saving in time in passing boats from one level to another and the small consumption of water as compared with common locks.

Green concluded by noting in his paper that the first ideas on this principle were justly due to the late Dr James Anderson of Edinburgh, who published a paper on the subject in his *Agricultural Survey of the County of Aberdeen* about the year 1793, but that it could be seen on a perusal of that able paper, that the details by which the principle was to be carried out were left much to the practical man.[36]

Green was not going to let precedents overshadow his contributions to engineering in producing a system based on Anderson's theory which was 30 years old!

4.9 NEWPORT DOCK – DISMISSAL BY DEVON QUARTER SESSIONS

A contract for the construction of a dock at Newport had been let in 1835, but within two years the contractors were in trouble and some time around 1840 James Green was appointed to take over from the previous resident engineer to complete the works. Once again he was far from Devon and at the Michaelmas Sessions of 1840, reported in the *Exeter Flying Post*,[37] Green was arraigned by one of the magistrates who said that Mr Green was renting a house in Wales and was in the services of the Newport Dock Company and that he had had no house in Exeter since Christmas 1838.

Mr Sillifant reported that Green had said that he could not continue his work in Devon satisfactorily without deputising the minor matters to his son. Some magistrates complained that they were having to do the work of the surveyor, and as we have seen from the Quarter Sessions minutes, Green was given 12 months' notice from Midsummer 1840.

Certainly the county needed the attention of an experienced engineer. Northbrook Bridge was being reconstructed, there were difficulties with the contract for Newton Poppleford Bridge, Laverton Bridge over the Yealm had been contracted for £96, for Cadworthy Bridge (Cadover) there was only one tender and it had been decided that Colyford Bridge had to be rebuilt. The severe reduction of Green's salary had misfired – he had just turned his full attention elsewhere, but that had not been successful either.

So Green left the county's employment and in 1841 was listed in the *Exeter Pocket Journal* as living in Heavitree with his son as 'Green, James and Son, Civil Engineers and Land Surveyors, Portview Cottage, Heavitree'.

The location of Portview Cottage appears to have been on the corner of Magdalen Road and Barrack Road, near Portview House. The June 1841 census listed the following as residents in Portview Cottage on census night.

Joseph D. Green   aged 30
Ruth Green   aged 20

At that time ages were rounded to the nearest five years.

James Green had already left the area, his notice having expired at the Midsummer Sessions of 1841 which were on 29 June, to concentrate on the rescue of the Newport Dock project. Joseph Green took up a post in Bristol in September 1843.

CHAPTER 5
1842-1849
WESTMINSTER, LONDON

5.1 INTRODUCTION

With Green's dismissal in 1841 the Midsummer Sessions appointed Thomas Whitaker as County Bridge Surveyor for Devon. Whitaker was the surveyor to the Exeter Commissioners of Improvement who had supervised the construction of the approach to the cast iron viaduct in North Street and the completion of this new structure. After his appointment Whitaker supervised the construction of Lord Rolle's bridge over the river Torridge by Castle Mills, the terminus of the Torrington Canal. He also replaced two of the arches of the Countess Wear Bridge over the river Exe with a single arch of 60 feet span thereby reducing the number of arches to six and giving the bridge an asymmetrical appearance.

Green with Rennie, the son of Green's employer of 1805, as contractor, brought the work at Newport Dock to a successful conclusion and then it was 'in 1843, too late in his career, that he settled in London, for owing to the active competition of younger men, he was not so extensively employed as he might have been.'[1] At the time he was 62 years of age and had done well to survive so long in a life that involved so much travelling.

Green turned his attention to more academic matters, giving evidence in Parliament on behalf of the Exeter City Council and then, in collaboration with Philip de la Garde, published an important paper on the history of the Exeter Canal. De la Garde was a surgeon, the author of a book on cataracts and member of the Devon and Exeter Archaeological Society, who had published many papers on archaeological subjects. He was also the last mayor of the unreformed corporation, as the Corporations Act of 1835 brought about election to a reformed council which admitted Non-conformists for the first time.[2,3]

While Green moved to London, his son Joseph took an appointment as superintendent of works at Bristol Docks at the end of September 1843. At this time I.K. Brunel had been the free-lance engineer to the dock company since 1832. Joseph Green's appointment was, in practice though not in name, the office of docks engineer but there was no question of him initiating substantial engineering work, or in any way supplanting Brunel.

It was during this time, 1846–47, that James Green was consulted by the city of Bristol on the sewerage of the river Frome and the floating dock. Through the Bristol Dock Act 1848, the dock company was taken over by the city of Bristol on 30 June 1848 and at about that time Brunel completed his last assignment for the dock company. James Green died in February 1849 and Joseph Green left Bristol in October 1851.

5.2 COMPLETION OF NEWPORT DOCK

The year 1835 heralded the incorporation of the Newport Dock Company which in turn was given sanction to build a five acre dock. The Harbour Act was passed on 21 June 1836. The membership of the Newport Dock committee was almost identical to that of the Monmouthshire Railway and Canal Company.

The first contract was let to Dike and Meyrick of Bristol in November 1835 and the work was expected to take two years. By 1837 it was clear that adequate progress was not being made and a letter to the *Monmouthshire Merlin* referred to the late contractors Dike and Meyrick and to the work of Mr William Armstrong of Bristol, the site engineer. In the autumn of 1837 Mr Cubitt was called in to give expert advice and confirmed that funds had not been misapplied and approved the siting, the plans and the work carried out to date.

In December 1840 it was reported that work on the dock had again been suspended in consequence of the failure of the contractor but that work would be resumed as soon as weather permitted. It was also stated that the dock company had not prepaid the contractor, so that no loss would be sustained. In February 1841 a revised estimate of £131,900 was submitted to complete the dock.

On 3 April 1841 the local press reported:

> For the fifth and we believe the last time the works of the Newport Dock are again resumed – the Government has advanced a sum which will enable the spirited projectors to accomplish their noble undertaking and before the end of the ensuing summer our Port will afford safe accommodation to vessels of the largest class.

Early in May 1841, a further general meeting was held in the Dock House. The engineer, James Green reported:

> Gentlemen. It is a source of pleasure to me to have to congratulate you on the improved position of this work since your last Half-Year General Meeting. The late Contractor having declared his inability to complete his engagements, no time was lost in taking the work out of his hands and taking possession of his implements and plant, and the favourable negotiations with the Exchequer Loan Board justifying the expectation of an advance of money, the plans and specifications for completing the work were prepared and advertisements issued for tenders.
>
> A contract has been entered into with Mr Rennie of Bath – a person of considerable experience in such works – for their entire completion, for a definite sum of money, by 25 March next year. Mr Rennie commenced his works on 22 March and since that time has been proceeding with much spirit, gradually increasing the number of men employed at the Dock to 300 besides 60 men and the necessary horses at the quarries.

Very satisfactory progress has been made with the Sea Lock, of which little now remains to be done. The second pair of lower gates is now completed and on their way from Shrewsbury and expected to arrive hourly.

I believe every necessary arrangement and preparation has been made for the completion of the works, to the extent contemplated and approved by Mr Walker, on behalf of the Exchequer Loan Commission, and I flatter myself that every proprietor in the undertaking must, on inspection of the works, be now satisfied in the prospect of their speedy and successful conclusion.

In the event the four and a half acre dock was opened on 10 October 1842 at a cost of £180,000. The main lock was 62 feet in width and 202 feet long, capable of berthing ships of 1500 tons displacement. There was in the lock a third pair of gates pointing towards the river to prevent high tides from flooding into the basin when it was found desirable to maintain the water in the dock at a uniform height to facilitate the loading of coals from the wharfs. It was proposed to maintain at all times, a depth of 18 feet in the basin and at such times there would be from 12 to 14 feet of fall from the wharfs to the water in the basin (an average in loading coals) and, whenever vessels of greater draught required it, the necessary depth of water could be furnished from the reservoir behind and north of the dock.

The basin or dock (walled on all sides) was 795 feet long, 240 feet wide and the depth of water might be varied from 18 to 30 feet. There was room for 12

Newport Dock 1842

vessels of the largest class to load or discharge at the same time and space in the centre for at least 30 vessels.

The reservoir for water north of the dock was of equal area and like form to the dock itself, with 11 feet depth of water. It communicated by a boat lock at the north end with the Monmouthshire canal, and a similar lock at the south end with the dock enabling vessels from the canal to berth alongside vessels in the dock. The tram roads from the hills and interior of the county had a direct and perfect communication with all wharfs in the dock.

The reservoir was designed to form a dock similar to the present one and its construction as an upper

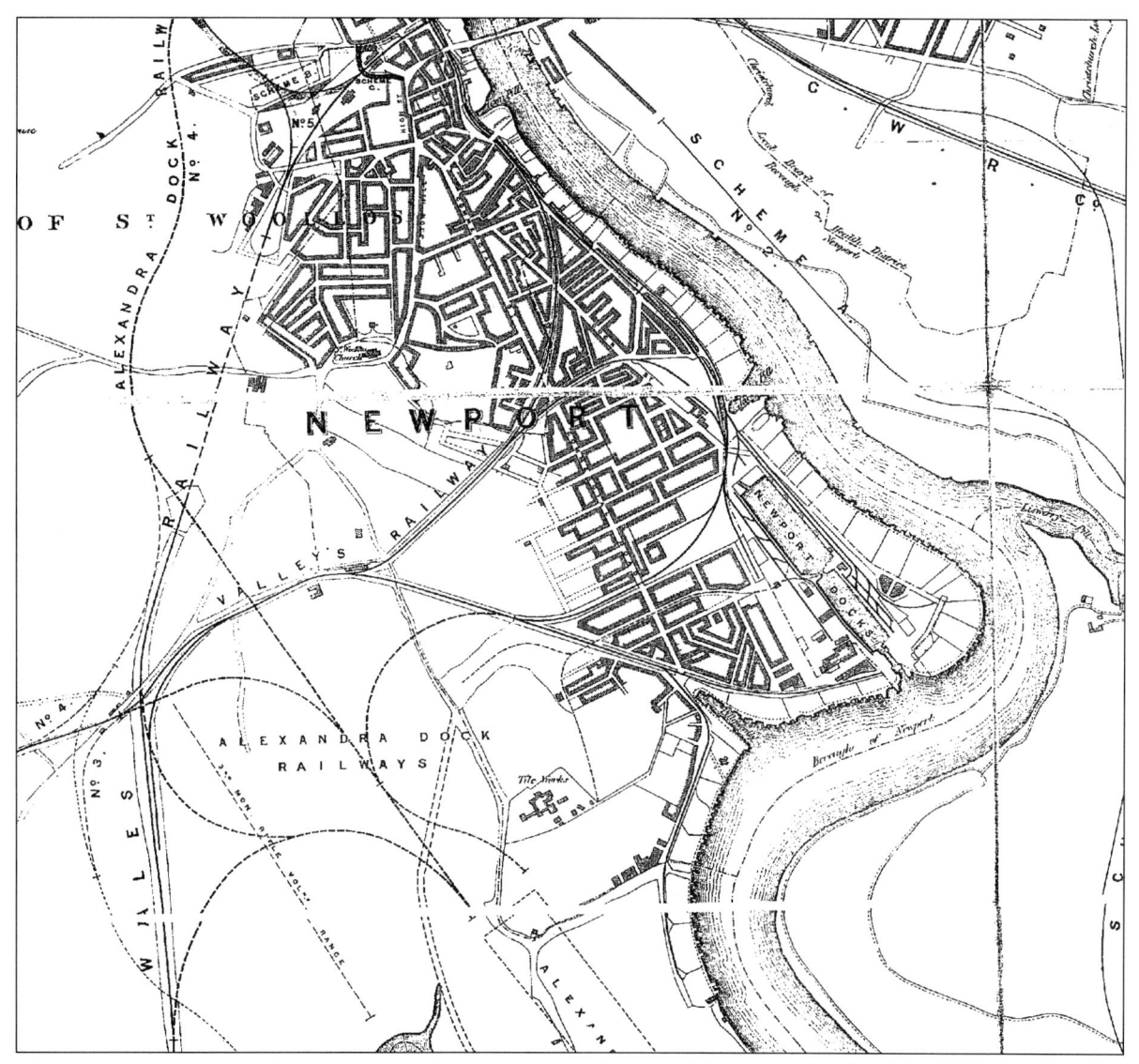

Plan of Newport Dock

dock could proceed without interruption to the trade of the lower dock. This upper dock would be best suited to general merchant vessels, leaving the lower dock for the coal and iron trades.[4, 5]

This was the last major civil engineering project supervised by James Green.

5.3 SOUTH DEVON RAILWAY BILL 1844

In 1844 Green was consulted about the building of the South Devon railway because of his knowledge of the estuary of the Exe.[6] Exeter City Council was opposing the Bill to safeguard its navigational rights in the Exe estuary and James Green made a report on 29 April 1844 to John Gidley, town clerk. He followed this by another letter on 1 May and gave evidence to a committee of the House of Commons (Sir John Mordaunt in the chair) on 2 May when he was examined on behalf of the South Devon railway by Mr Rogers.

The essence of Green's evidence was that the embankment alongside the estuary would enclose 102 acres which would make a significant difference to the movement, and hence the scour, of the water of the estuary as it crossed the bar. Thus, with reduced movement, less scour would occur. He referred to his work at Saltram for Lord Morley and to an embankment at Exmouth for Mr Hill.

Under cross-examination he admitted that his views had changed over the forty years since he carried out his work at Saltram and denied that his walls had been breached by the tides. He suggested that a solution would be to place the railway on a piled structure to allow the ebb and flow of the tide to have full effect.

Examination of the 1809 and 1972 1" OS maps shows that, in the event, the South Devon railway was built against the edge of the estuary (and the highway) from just north of Powderham church to Langstone Rock, with the exception of Cockwood harbour and a tidal entry north of the Warren that reached as far as Shutterton Bridge (A379). Most of the 102 acres would have been in these places. Today Cockwood harbour is open because it is spanned by two bridges, whereas the tidal reach to Shutterton Bridge has been restricted by a tidal gate at the railway allowing the inland area to be developed for caravans. Apart from this latter area everything seems to have been done by the South Devon railway to avoid enclosure or encroachment on to tidal areas.

5.4 PAPER ON THE EXETER CANAL TO THE INSTITUTION OF CIVIL ENGINEERS

Paper No. 671 titled *Memoir of the Canal of Exeter from 1563 to 1724* was written by Philip Chilwell de la Garde; with a continuation by James Green M Inst C E.[7]

Mr de la Garde's paper was of first importance. It had originally been undertaken with a view to a communication to the Society of Antiquaries but on being requested by Mr Green to give the Institution of Civil Engineers the result of his researches into the engineering portion of the history, Mr de la Garde readily complied. It was a very interesting narrative as it fixed the date of the introduction of pound locks into the country.

Plate I of the paper was prepared by J. Green Jnr and was a plan of the Exeter Canal in the reign of Queen Elizabeth. It showed the mill leat to Wear Mills, which Trew had originally proposed in 1563, converting into a canal and a short length of tributary of the river Exe well upstream of the mills on the other bank which was probably the original course of the Alphin brook. For the construction of Trew's canal, the Alphin brook appears to have been diverted downstream parallel to the canal until it joined the Matford brook and here the two rivers formed the deep pyll which was joined by the canal.

Three locks were situated on the canal, one a furlong upstream of the pyll, another on the site of the present double locks and a third a quarter of a mile downstream of King's Arms Sluice. De la Garde discussed at length the size of these locks which were pools up to 300 feet in length and 80 feet in breadth between two gates and capable of receiving several vessels at the same time. He said that 'the waste of water was very great, but the engineer had the river Exe at his disposal.'

Green's paper No. 671A included a large scale map of the canal to Turf at a scale very close to 1/25,000 with a longitudinal section of the canal.

He described how he overcame the difficulties of excavating for the lock at Turf, which was a considerable feat of civil engineering. A summary of his reports of October 1820, 1 March 1824 and 24 March 1826 is given, together with Telford's reports on Green's advice dated 31 July 1824 and 23 June 1826.

Both de la Garde and Green's papers are of considerable interest today as they give an account of the early work of Trew and show how the city monitored Green's enthusiastic proposals by obtaining advice from Telford.

5.5 JOSEPH DAND GREEN

In 1843 the Bristol Dock Company's salaried staff consisted of a couple of clerks, the senior being the equivalent of a modern company secretary, and a Dockmaster who controlled the shipping and supervised the harbour with the assistance of some 15 lockmen and labourers. The committee decided to reform its administration to have a clear distinction between the functions of Dockmaster and a new officer to be called the superintendent of works. The duties of the latter were to be those of a resident engineer but this title was rejected, possibly to avoid confusion with Brunel and his assistants. Brunel had been retained since 1832 to advise the company.

The new posts were advertised and filled at the end of September 1843, the new superintendent of works being Joseph Dand Green. He quickly got to work and within nine days had put in his first recommendation to the Board. By the end of the first month he had prepared a detailed report of his field of work. He inspected the masonry of the harbour, investigated the condition of the towing paths and was allowed to keep a horse on expenses. In December 1845 he submitted plans for new workshops and the following year reported that they had saved the company much money. To begin with there was no question of him initiating engineering work but with the withdrawal of Brunel in 1848 he seems to have expected more responsibility. Under the Bristol Docks Act 1848 the dock company passed the docks to the city of Bristol on 30 July 1848. Finding he was not obtaining this responsibility, Joseph Green resigned on 20 October 1851 giving as his reason the complaint that the commissions which should have come to him were instead given to an independent engineer.[8]

In 1856 Joseph Green was working on the New River Project at Wisbech, employed by James Rendel. Rendel wrote a letter to his son, George, who was then working for Sir William Armstrong. It was one month before James Rendel's death on 21 November 1856.

> 18 October 1856
> ... Mr Peneston has left – he sailed from England on Wednesday. This greatly added to my perplexity just now for things are in a sad mess and to get them right will demand no small share of my time and patience.
>
> Peneston's successor is Mr Green, son of old Green who was long County Surveyor and great engineer of Devon and the West Country. He is a plodding careful fellow, not so bright but will I hope answer our purpose. He has no family and has therefore taken lodging at Wisbech which Dent had, Dent having asked me to allow him to go to Pernambuco with Peneston.[9]

So whereas at the Epiphany Quarter Sessions of 1831 Rendel had been prepared to undercut James Green for the post of County Surveyor here he was, 25 years later, giving employment to James' son Joseph, recognising perhaps that Joseph needed help.

5.6 THE DRAINAGE AND SEWERAGE OF BRISTOL

When the floating harbour at Bristol was formed by placing locks on the river downstream of the docks and diverting the river Avon along a new channel to the tideway below the locks, no thought was given to intercepting and carrying off the sewage of the city away from the harbour. Further sewage was brought in by the tributary river Frome which passed through a populous part of the city.

In 1846 Green was instructed by the council to advise on the measures necessary for abating the nuisance and he reported on 30 October. His report recommended straightening the river Frome, making it of uniform width to give the greatest scour of the bed and intercepting the sewers that discharged into it. The council considered that it should not proceed because it did not have the necessary legal powers but further instructed Green in January 1847 to advise on action to be taken between Stone Bridge and Castle Moat. The report was made on 16 March 1847 and, during the summer, works were carried out at a cost of £4537 to clear this area of accumulated sludge.

Green presented a paper on these reports and the work carried out to the Institution of Civil Engineers on 8 February 1848 when the discussion was initiated by the Dean of Westminster and extended over three evenings.[10]

5.7 HIS DEATH

James Green died on 13 February 1849 from a heart attack at 67, Manchester Buildings, Westminster. He was in his sixty-eighth year. He was buried on 20 February at Bunhill Fields as a non-member of the Religious Society of Friends, though his connection was enough for a Quaker burial.[11]

His obituary was published in the *Proceedings of the Institution of Civil Engineers* in 1850 and this was noted by the *Bristol Mirror* who commented that his son Joseph D. Green was resident engineer at Bristol Docks.[12]

5.8 A REVIEW OF GREEN'S ACHIEVEMENTS

Green achieved a double reputation as a civil engineer, widely acclaimed as a canal engineer yet producing an impressive list of bridges for the county of Devon.

As their bridge engineer he was able to replace many of the original 236 bridges with structures that have endured to this day on the important roads of Devon. Fenny Bridge, for instance, carries daily 15,000 vehicles, with weights up to 38 tons and with exceptional loads as well. The fact that the bridge measures 20 feet between parapets means that it is just wide enough to take two way traffic, but presents a great risk to pedestrians. Much the same can be said of Cowley Bridge. Only Thorverton Bridge and the cast iron spans have become unserviceable, the first because of the approach inclines in horse-drawn days, the latter because, like many new materials, cast iron was an unsuitable material for the heavy loads that now have to be carried. Cast iron did prevail on Barnstaple Bridge where pedestrian weight remained constant and the design served its purpose until no longer needed. The reader can best assess the quality of Green's bridges by visiting the river crossings and seeing them for himself.

As a canal engineer, Green served Exeter well and elsewhere the tub-boat canals were well suitable for the traffic and topography of Devon. Unfortunately, the canals were built only a few years before the railway began to show its superiority. The inclined planes were undoubtedly a clever solution to the problems of ascending the Devon hills. He had more lifts working for longer than elsewhere in Britain, and both these devices would have been better if Green had paid more attention to detail and not tried to be contractor as well.

In taking the role of contractor as well as county bridge engineer and canal consultant, Green may well have made some good profits as first but he quickly sustained losses working for the county so that this practice was ended within three years. Of course, we do not know how his finances stood after constructing the embankments for Lord Boringdon and the widening of Bideford Bridge, but he was soon able to build a magnificent house in Exeter at Elmfield and carry out other significant architectural work for Devon gentlemen so all must have been well. The Braunton scheme and St

David's church may have provided him with the money to put £3000 into the Bude Canal scheme but thereafter that project may not have helped him financially. Through the 1820s he was earning fees for the Exeter Canal and his advice to the various turnpike trustees, but the latter fees were not large. Nevertheless, £60 for the designs of the two bridges for the Barnstaple trustees was a useful sum when added to his other income.

What seems certain is that he over-reacted to the reduction in his salary from the county in 1830. From £550 to £300 was a large drop but to take on work as far apart as South Wales, Taunton and the London–Birmingham canal was ridiculous in an age when travel was by horse or ship. Yet in 1833 this is what he was doing and the result was catastrophic with bankruptcy in 1837 because, once again, he had been tempted to contract to complete the works in Wales for which he was also engineer.

Green must have been unique among County Surveyors in going bankrupt and the effect on this man of 56 years must have been serious indeed. Yet in those latter years excellent county bridges were still being built. So his son Joseph must have gained useful experience which he used to become an Associate of the Institution of Civil Engineers at an early age.

There appears to be no record of consulting works between 1837 and 1840 though James Green may have arrived at Newport Dock earlier than 1840. When he completed this project and moved to Westminster it seems that he retained useful contacts in Exeter. This is made evident by his work on the South Devon Railway Bill and his paper in conjunction with de la Garde.

Green's work on the sewerage of Bristol must have seemed relevant to London since it induced the Dean of Westminster to introduce the discussion of his paper and for the discussion to last for three evenings. So despite his reduced workload, James Green appears to have remained an effective civil engineer, if a controversial one, right up to his sixty-eighth year, and this was no mean achievement.

To have completed one of the early significant dams, a very large mansion, a church, canals, bridges, the realignment of long lengths of the turnpike road and the reclamation of land was an achievement that brings Green's name to the attention of any researcher of south western civil engineering from 1806 to 1841, and this is his enduring title to fame.

APPENDICES

APPENDIX A

THE REPORT OF JAMES GREEN ON THE DEVON COUNTY BRIDGES – 10 JANUARY 1809

Green produced an imperial sized book of handwritten entries for each bridge with two or three entries to a page. Each entry gave the name and location, the span(s) and the width between parapets. The method and material of construction were described and each entry was completed by a report on the condition of the structure.

The entries were made in the order in which they were inspected. Green started with Stoke Canon and went west via Jacobstowe to Tamarstone, thence via Werrington to Lifton, Tavistock, across Dartmoor to Dartmeet, then back to Meavy and Plym, Lee Mill and Ivybridge. Another trip began at Cowley and extended to Yeoford and back to Codshead pool.

Then again, he went from Exe Bridge, Morebath, to Uffculme, Crediton and Ellerhayes, near Killerton. The next trip was from Honiton Clyst to Cadhay, Yarty, Bow at Axminster, the Axe Bridges at Colyford and back to Sidford and Heavitree. A short trip was made from Withy to Clyst St Lawrence. His next itinerary covered the route from Yealmpton to Ermington, Loddiswell, Kingsbridge, Blackawton, Ashprington, Harberton, Littlehempston, Staverton, South Brent, Buckfastleigh and Kenn.

The next excursion was to Chudleigh, Bridford, Moretonhampstead, North Bovey, Bovey Tracey, Bickington, Ashburton, Newton Bushell, Abbots Kerswell, Kingsteignton, Ashcombe and John Tolls. Lastly he went to Clapper Bridge at Honiton.

Green must have been an excellent horseman and very fit to cover this ground between Michaelmas and Epiphany. Whether he kept his book at home or wrote up his notes in his lodgings is unclear but he provided an alphabetical index at the end of the book.

Each bridge is numbered. Although he finished with number 236, both Tackett Wood and Collapit are numbered 145, so there are 237 entries. But there are separate entries for Long Bridge, Plympton, (69) and New Bridge, Plympton (70). He notes that New Bridge was found on enquiry to belong to the trustees of the Plymouth Eastern Turnpike. Unfortunately these appear to be the same bridge, for the Ordnance Survey marks the site of the Plym crossing as New Bridge while in 1833 a contribution was being made by the county to the trustees for Long Bridge and the resulting bridge in Green's style is very much over the Plym! Today the county calls the site Long Bridge.

There are two bridges in Kingsteignton called New Bridge, with greatly different dimensions recorded. New Bridge is marked on the 1809 Ordnance Survey and today's Ordnance Survey as crossing the river Teign so this would appear to be the Kingsteignton/Ilsington Bridge number 211. The site for Bridge 221 may be that on the approach to 211, 50 yards to the south, now named Rixley.

The itinerary numbers and names are reproduced herewith. In the following list the bridges have been taken in the order of their first initial and assigned a map reference, span(s) and width between parapets. Every map reference cannot be guaranteed, because while most are easily identified, some have names not currently in use in the county and research with older maps has not always been conclusive.

An examination of the report shows that only nine of the 236 bridges had a span or spans greater than 30 feet. The nine bridges were Exe Bridge, Morebath, with spans 25, 30 and 25 feet; Fenny Bridge, which Green had just rebuilt in 1808 with spans of $41^1/_2$, 47 and $41^1/_2$ feet; Lee Mill, also just rebuilt by Green in 1808 with a span of 36 feet at 21 degrees skew; Last Bridge, Cullompton, 30 feet; Pynes Bridge, Upton Pyne, at 30, 16, and $9^1/_2$ feet; Thorverton Bridge (1796) over the river Exe at 30, 40 and 30 feet; New Bridge, Tawstock, then being rebuilt at 30, 40 and 30 feet; Tarr Bridge over the Yealm (1797), spanning 30 feet; and Gara Bridge, Diptford and Halwill, all 30 feet span. 27 bridges had a width between parapets greater than 12 feet, but 110 had a width of only 9 feet or less.

THE LIST OF DEVON COUNTY BRIDGES 1809 IN ITINERARY ORDER
(Spelling as in original documents)

1   Stoke Canon
2   Thorverton
3   Higher Creedy
4   Bow in Bow
5   Bunty in Bunty
6   Taw, North Tawton
7   Newland, North Tawton
8   Sticklepath, South Tawton
9   Brightly, Okehampton
10 Jacobstowe, Jacobstowe
11 Hele, Hatherleigh
12 New, Huish
13 Beaford, Beaford
14 Hatherleigh, Hatherleigh
15 North Lew, North Lew
16 Black Torrington, Black Torrington
17 Dipford, or Diphard Mill
18 Bridgerule, Bridgerule
19 Tamarstone, Pancrasweek
20 Helscott, North Petherwin
21 Yeolm, Werrington
22 Druxham, Werrington
23 New, Lifton
24 Hele, Lifton
25 Poulston, Lifton
26 Lifton, Lifton
27 Southern, Lifton
28 Tinney, Lifton
29 Stowford, Stowford
30 Hayne, Stowford
31 Wrixele, Thrushelton
32 Combebow, Bridestowe
33 Lidford, Lidford
34 Coryton, Coryton
35 Kidnole, Milton Abbot
36 Marystow, Marystow
37 Sydenham, Marystow
38 Griston, Bradstone
39 Horse, South Sydenham
40 New, Tavistock
41 Middle Lumburn, Tavistock
42 Higher Lumburn, Tavistock
43 Lower Lumburn, Tavistock
44 Shilly Mill, Tavistock
45 Denham, Buckland Monachorum
46 Sticklepath, Buckland Monachorum
47 West, Tavistock
48 Abbey, Tavistock
49 Harford, Mary Tavy
50 Lower Hill, Peter Tavy
51 Higher Hill, Peter Tavy
52 Merrivale, Whitchurch
53 Hockworthy, Sampford Spiney
54 Ward, Walkhampton

Appendices

55	Two Bridges, Withecombe on the Moor	98	Culmstock, Culmstock
56	Belliver, Withecombe on the Moor	99	Long, Cullompton
57	Dartmeet, Withecombe on the Moor	100	Stoniford, Cullompton
58	Hexworthy, Withecombe on the Moor	101	Last, Cullompton
59	Norsworthy, Walkhampton	102	Etherhay, Broadclyst & Silverton
60	Sheepstor, Sheepstor	103	Honiton Clyst, Clyst Honiton
61	Harrowbridge, Buckland Monachorum	104	Rockbeare, Rockbeare
62	Hew Meavy, Sheepstor	105	St Saviours, Ottery St Mary
63	Lower Meavy, Sheepstor	106	Dead, Ottery St Mary
64	Higher Meavy, Sheepstor	107	Cadhay, Ottery St Mary
65	Cadworthy	108	Corner Cut, Ottery St Mary
66	Shaugh, Shaugh	109	Gosford, Ottery St Mary
67	Bickleigh, Shaugh & Meavy	110	Finney, Feniton
68	Plym, Plympton St Mary	111	Town, Honiton
69	Long, Plympton & Egg Buckland	112	Tailwater, Talaton/Ottery St Mary
70	New, Plympton	113	Yarty, Kilmington
71	Lee Mill	114	Bow, Axminster
72	Parsonage, Cornwood	115	Stoney, Axminster
73	Langham, Cornwood	116	Weycroft, Axminster
74	Hartford, Hartford	117	Ax, Axmouth
75	Ivybridge, Ermington & Ugbrook	118	New, Axmouth
76	Cowley, Brampford Speke	119	Bridge adjoining Ax Bridge
77	Pynes, Upton Pyne	120	Cullyford, Culliton
78	New, Upton Pyne	121	Sidford, Sidbury
79	Shuttern, Newton St Cyres	122	Newton Poppleford, Newton Poppleford
80	Lower Creedy, Newton St Cyres	123	Otterton, Otterton
81	Long, Newton St Cyres	124	Grindle, Clyst St Mary
82	Little Silver, Shobrooke	125	Bishop's Clyst, Sowton
83	Priston, Sandford	126	Heavitree, Heavitree
84	Heath, Sandford	127	Withy, Broadclyst
85	Praydon, Crediton	128	Burrow, Broadclyst
86	Dowridge, Sandford	129	Ashclyst, Broadclyst
87	Yeoford, Crediton	130	Lawrence Clyst, Lawrence Clyst
88	Neopardy, Crediton	131	Calves, Morchard Bishop
89	Gunstone, Crediton	132	New, Tavistock
90	Meetford, Crediton	133	Puslinch, Yealmpton
91	Uton, Crediton	134	Tarr, Yealmpton
92	Culverly, Crediton	135	Yealm, Yealmpton
93	Fordton, Crediton	136	Laverton, Yealmpton
94	Codshead Pool, Crediton	137	Ermington, Ermington
95	Exe, Morebath	138	Little Ermington, Ermington
96	Five Bridges, Cullompton	139	Sequers, Ermington
97	Uffculme, Uffculme	140	Goulsford, Modbury

#	Bridge, Parish	#	Bridge, Parish
141	Aveton Gifford, Aveton Gifford	183	Dart, Buckfastleigh
142	Hatch, Loddiswell	184	Holn, Holn
143	Loddiswell, Loddiswell	185	New, Holn
144	Topsham, Loddiswell	186	Buckland, Buckland-on-the-Moor
145	Tacket Wood, West Allington	187	Ponsworthy, Buckland-on-the-Moor
145	Collapit, West Allington	188	Cockmanford, Withecombe
146	Blanks Mill, Marlborough	189	Kenford, Kenn
147	Bocombe, Sherford & Buckland	190	Kately, Chudleigh
148	Ranscombe, Sherford	191	Eightly, Chudleigh
149	Frogmore, Sherford	192	Chudleigh, Chudleigh
150	Bow, Blackawton	193	Crocombe, Hennock
151	Slapton, Slapton	194	Bramble, Trusham
152	Blackpool, Blackawton	195	Spara, Ashton
153	Old Mill, Dittisham	196	Christow, Christow
154	Tuckenway, Ashprington	197	Bridford, Bridford
155	Bow, Ashprington	198	Steps, Dunsford
156	Washburn, Halwill	199	Clifford, Drewsteignton
157	Gerah, Diptford & Halwill	200	Fingle, Moretonhampstead
158	Bickham, Diptford	201	Doggermarsh, Chagford
159	Harbertonford, Harberton	202	Rushford, Chagford
160	Roster, Harberton	203	Chagford, Chagford
161	Lee, Harberton	204	Ash, Throwleigh
162	Yeo, Harberton	205	Paynes, Throwleigh
163	Pords, Stoke Gabriel	206	Broad, North Bovey
164	Popes or Tukeway, Cockington & Paignton	207	South, North Bovey
165	Gatcombe, Littlehempston	208	Drakeford, Listleigh
166	Fishacre, Littlehempston	209	Wolford, South Bovey
167	Little Forder, Littlehempston	210	Bovey, South Bovey
168	Great Forder, Littlehempston	211	New, Kingsteignton & Ilsington
169	Littlehempston, Littlehempston	212	Great Jews, South Bovey
170	Shinners, Dartington	213	Little Jews, South Bovey
171	Staverton, Staverton	214	Drumbridge, South Bovey
172	Austin's, Staverton	215	Bow, Bickington
173	Damerels, Buckfastleigh	216	Chuley, Ashburton
174	Church, Buckfastleigh	217	Gulwell, Ashburton
175	Leedy, South Brent	218	Kings, Ashburton
176	Shipley, South Brent	219	Great, Ashburton
177	Ball, Buckfastleigh	220	Chipley, Bickleigh
178	Dean Burn, Dean Prior	221	New, Kingsteignton
179	Brent Harbertonford, South Brent	222	Chickham, East Ogwell
180	Glaze, South Brent	223	Morely, Torbryan
181	Brent, South Brent	224	Tuckers, Highweek & Newton Bushell
182	New, South Brent	225	Lemon, Newton Bushell

226 Keyberry, Combe-in-Teignhead
227 Longford, Abbots Kerswell
228 Aller, Abbots Kerswell
229 Dock, Kingskerswell
230 Bow, Ipplepen
231 Ventover, Teigngrace
232 Dawlish, Dawlish
233 Teign, Kingsteignton
234 Ashcombe, Ashcombe
235 John Tolls, Kenton
236 Clapper, Honiton & Combe Raleigh

THE BRIDGES IN THE 1809 LIST PLACED IN ORDER OF THEIR FIRST INITIAL WITH LOCATION, SPANS AND WIDTHS

(Original spelling given with some modern names in capitals)

NAME	LOCATION (Nat. Grid Ref.)	SPAN(S) Feet	WIDTH Feet
Abbey, Tavistock	SX482743	2x13	15$^1/_2$
Axe, Axmouth	SY259926	21, 23, 21	9
Axmouth, New	SY259926	10	12
Ashclyst, Broadclyst	SY009982	14	10
Aveton Gifford	SX692470	6x16, 16, 3x12	12
Austin's, Staverton	SX750660	3x21, 2x16	7$^1/_2$
Ash, Throwleigh Moor	SX677905	8	9
Aller, Abbotskerswell	SX876689	14	10$^1/_2$
Ashcombe	SX913795	12	9
Ashburton, Great Bridge	SX754702	13	9$^1/_2$
Bow, in Bow	SS717017	18, 15	8
Bunty (Bundleigh)?	SS657045	2x16$^1/_2$	8$^1/_2$
Brightley, Okehampton	SX598974	3x14	6$^1/_2$
Beaford	SS542143	20, 28, 20	8$^1/_2$
Black Torrington	SS469060	2x27	10
Bridgerule	SS274028	2x11	10
Bickleigh, Shaugh	SX526618	24	7
Bow, Axminster	SY290982	15, 22, 15	20
Bridge, 40ft east of Axebridge	SY260926	17	7
Bishop's Clyst, Sowton causeway	SX970911	2x15,8,(2x11)	9$^1/_2$, (11)
Burrow, Broadclyst	SX983975	17	10
Blank's Mill, Malborough	SX726410	5,4	11
Bocombe, Sherford & Buckland	SX750443	5, 5	9
Bow, Blackawton	SX807493	10	8
Blackpool, Blackawton	SX852478	8	10
Bow, Ashprington	SX812565	17	9$^1/_2$
Ball, Buckfastleigh (HIGHER TOWN)	SX736663	12	7$^1/_2$
Brent, Harbertonford	SX717623	4x4	F.P.
Brent, South Brent	SX697595	26,14	13
Buckland (on Moor)	SX718719	20	8
Bramble, Trusham	SX849812	11	7
Bridford	SX834872	2x22	9$^1/_2$
Broad, North Bovey	SX721848	12, 1	8$^1/_2$
Bickham, Diptford	SX725553	2x12	11
Bovey, South Bovey	SX814782	2x16, 11	10
Bow, Bickington	SX795724	14	12
Bow, Ipplepen	SX835650	9	9
Bellever	SX658773	8	F.P.
Creedy, Higher; Crediton	SX846012	18, 8, 6	10$^1/_2$
Cambrow, Bridestowe (COMBEBOW)	SX485879	2x10$^1/_2$	9$^1/_2$
Coryton	SX441833	2x12	10
Cadworthy, Sheepstor & Meavy (CADOVER)	SX555646	2x11	6$^1/_4$
Cornwood, Parsonage	SX614603	2x11	10
Cowley	SX907955	4x21	9$^1/_2$
Creedy, Lower; Newton St Cyres	SX880986	2x20	12
Culvery, Crediton	SX834985	14, 2 small arches	8$^1/_2$
Codshead Pool, Crediton	SX856991	4x8	12
Culmstock	ST101137	3x20, 2x16, 10	10
Clyst Honiton	SX985935	3x13$^1/_2$, 12$^1/_2$	13
Cadhay	SY093960	14, 20, 14	8$^1/_2$
Corner Cut, Ottery St Mary	SY092960	Wooden F.B. now extinct	
Clapper, Combe Raleigh	ST163011	2x12, flood arch 16	

157

Cullyford, Colliton	SY254926	2x15	10$^1/_2$	(FENNEY)	SY114985	41$^1/_2$, 47, 41$^1/_2$	20
Calves	SS749087	2x12	14	Frogmore, Sherford	SX775426	4, 2	9
Collopit, West Allington	SX728421	8	11	Fishacre, Littlehempston	SX818644	3x8	9
Church, Buckfastleigh	SX742662	14	20	Forder, Little			
Cockmanford,				(LONGFORD)	SX812642	10$^1/_2$, 4$^1/_2$	8$^1/_2$
Widecombe	SX717751	7	7$^1/_2$	Forder, Great (FORD)	SX815638	12$^1/_2$	10
Chudleigh	SX857784	2x27, 16	10	Fingle	SX743899	3x16$^1/_2$	6$^1/_2$
Crocombe, Hennock	SX848811	2x18, 22	6$^1/_2$				
Christow	SX839866	2x20	7$^1/_2$	Griston, Bradstone			
Clifford	SX780897	3x16$^1/_2$	7$^1/_2$	(GREYSTON)	SX368803	5x22, 14, 10	10
Chagford	SX693879	3x17	7$^1/_2$	Gunstone, Crediton	SX805985	2x12, 6	8$^1/_2$
Chuley	SX750692	6	9$^1/_2$	Gosford	SY101970	20, 23, 20	8$^1/_2$
Chipley, Bickington	SX810719	12	8$^1/_2$	Grindle	SX974903	2x4, 3x8, 2x4	15
Chickham, East Ogwell	SX833710	14	9	Goulsford, Modbury	SX636515	12	18
				Gatcombe,			
Dipford Mill	SS438065	3x12$^1/_2$	10	Littlehempston	SX818625	2x5	14
Druxhan (DRUXTON)	SX345884	2x12, 2x10	8$^1/_2$	Glaze	SX690590	14	10
Denham,				Gulwell, Ashburton	SX753693	13	9$^1/_2$
Buckland Monachorum	SX477678	27, 12	9$^1/_2$	Gerah, Diptford (GARA)	SX728534	30	9$^1/_2$
Dartmeet, Widecombe	SX671731	2x26	15				
Dowridge, Sandford				Hele, Hatherleigh	SS540063	3x20, 15	10
(DOWRICH)	SS827046	14	7$^1/_2$	Huish, New	SS548112	2x19	8
Dead, Ottery St Mary	Site demolished	20	9	Hatherleigh	SS541042	3x13	10$^1/_2$
Damerell, Buckfastleigh	SX741661	14	9	Helscott, N. Petherwin			
Deanburn	SX732651	20	13	(Cornwall)	SS285877	2x9	6
Dart	SX744667	3x24, 2x22	8$^1/_2$	Hele, Lifton	SX359863	18, 11	9$^1/_2$
Doggermarsh	SX713893	3x17	7$^1/_2$	Hayne, Stowford	SX416867	4x8$^1/_2$	9
Drakeford	SX789801	14,12	8$^1/_2$	Horse, South Sydenham	SX400748	7x20	12
Drumbridge	SX829753	10	12	Harford, Mary Tavy	SX505767	2x18, 20	11
Dock, Abbotskerswell				Hill Bridge, Higher,			
(DACCA)	SX878677	10, 3	9	Peter Tavy	SX532804	4x7	Too narrow for wheeled carriages
Dawlish	SX954768	2x15	11				
				Hill Bridge, Lower,			
Exe, Morebath	SS929244	25, 30, 25	16	Peter Tavy	SX527795	2x5	Too narrow for wheeled carriages
Ellerhayes	SS975011	4x13$^1/_2$	15				
Ermington	SX640530	2x16	6$^1/_2$	Hockworthy,			
Ermington, Little	SX641530	8	7$^1/_2$	Sampford Spinney	SX531705	24, 12	7$^1/_2$
Eightly, Chudleigh				Hexworthy, Widecombe	SX658728	24, 18, 12	9
(HEIGHTLY)	SX859785	12	9	Harrow, Buckland			
				Monachorum (HORRA)	SX513699	2x13$^1/_2$, 11$^1/_2$	11
Fordton, Crediton	SX839990	2x13, 2x12	9	Hew Meavy, Sheepstor			
Five, Cullompton	ST026095	20	22	(HOO MEAVY)	SX526656	26	8
Finney, Feniton				Hartford, Harford	SX636595	21, 9	8$^1/_2$

Appendices

Heath, Sandford	SS844045	14	9		Lee, Harberton			
Honiton Town	ST159005	15	3/24/3		(BEENLEIGH)	SX798566	2x6	10
Heavitree	SX947921	2x10	22		Leedy, South Brent			
Hatch, Loddiswell	SX714472	2x15	6½		(LYDIA)	SX696607	20	9
Harbertonford	SX784561	2x9, 8	17		Lemon, Newton Bushell	SX860713	18	12
Hempston, Little	SX813625	2x16	9		Langford, Abbotskerswell	SX871690	10, 2	8
Holn (HOLNE)	SX730706	10½, 2x18, 27	9					
Holn, New	SX711708	23, 24, 13	8		Marystow	SX434832	15	8½
					Merrivale	SX550751	2x8	12
Ivy, Ermington	SX636563	22	11		Meavy, Lower	SX538671	23	8
					Meavy, Higher	SX546699	21	9½
Jacobstowe	SS592018	20½	12		Meetford, Crediton	SX817965	2x9 (timber)	8
Jews, Great	SX838764	2x20, 18½	10		Morely, Torbryan	SX827712	14	6
Jews, Little	SX837762	18	15					
John Tolls, Kenton	SX953846	21, 13, 3	12		New, Kingsteignton			
Kidnole,					(RIXLEY)	SX849763	13	9
Milton Abbot (COMBE)	SX412787	2x8½	8½		Newpardy	SX797986	3x12	8
Kenford, Kenn	SX915864	12	8½		North Lew	SX507991	13	8
Kateley, Chudleigh					Norsworthy, Walkham	SX567693	14	6
(BRIDGE LANE)	SX871793	14	10		Newland, North Tawton	SS658004	2x17	8½
Kingsteignton, New	SX849764	2x16, 10½	9		Newton Poppleford	SY079852	5x18	8½
Kings, Ashburton	SX755699	10	12					
Keyberry, Combe in					Otterton	SY079852	3x18	7½
Teignhead (PENN INN)	SX870707	3x8, 2x6	8		Old Mill, Dittisham	SX861519	2x6	8
Lifton, New (over Tamar)	SX348867	3x25, 5	12		Poulson, Lifton	SX355849	3x17	9½
Lifton, Lifton	SX389848	2x15,16	10		Plym, Plympton	SX523586	2x16	8
Lifton, Southern	SX396849	3x12, 5½	9		Priston, Sandford	SX849048	14	7 no ppts
Lydford, Lydford	SX509845	16	12		Praydon, Crediton			
Lumburn, Middle	SX454742	11	8		(PRIORTON)	SS833041	14, 8	8
Lumburn, Higher	SX452746	3x6	8½		Puslinch, Yealmpton	SX570510	2x13, 12	9½
Lumburn, Lower	SX459730	2x7	14		Pords, Stoke Gabriel	SX857576	10½	7½
Long, Plympton	SX519567	2x14, 2x12,11	11		Popes, Cockington	SX900625	7	10
Lee Mill	SX600557	36 (skew 21°)	20		Ponsworthy	SX702739	20	8
Langham, Cornwood	SX608591	2x11	12		Paynes, Throwleigh	SX658915	8	7
Long, Newton St Cyres	SX881987	2x15	12		Pynes, Upton Pyne	SX906959	30, 16, 9½	10
Little Silver, Shobrook	SS871014	14	10					
Long, Cullompton	ST025077	16, 18, 16	12		Rockbeare	SY009952	15	16
Last, Cullompton	ST027066	30	18		Rancombe, Sherford	SX757449	4½	9
Lawrence Clyst	ST027000	2x9	9½		Roster, Harberton	SX771563	2x10	10
Laverton, Yealmpton					Rushford, Chagford	SX705882	2x21	7
(LOTHERN)	SX596538	13, 15	10					
Loddiswell	SX719478	2x18	18		Stoke Canon	SX938975	3x14	10½

159

Sticklepath, South Tawton	SX643940	2x17	14	Tailwater, Talaton	SY092960	18	20
Stowford	SX427873	17	8	Tawstock, New	SS570282	30, 40, 30	18
Sydenham, Marystow	SX428838	2x13	8½	Tarr, Yealm	SX579515	30	9
Shilley Mill, Tavistock	SX466719	14	14	Topsham, Loddiswell	SX732511	20	10
Sticklepath, B. Mon'ch'm (GRENOFEN)	SX490709	16	7	Tackett Wood, W. Allington	SX734436	2x2½	14
Sheepstor (Now under Burrator Reservoir)		20	9	Tuckway, Ashprington (TUCKENHAY)	SX818560	7	9
Shaugh	SX523636	2x14½	7	Tuckers, Highweek	SX857712	11	26
Shuttern	SX881979	10	5	Teign, Kingsteignton	SX859734	2x18, 16, 16, 13, 12x9 9 9 9 14 7	
Stoneyford, Cullompton	ST029074	22	12				
St Saviours, Ottery St Mary	SY093951	2x27	10				
Stoney, Axminster	SY294987	2x20	16	Upton Pyne, New	SX900967	2x24, 14	12
Sidford	SY137899	15, 20	14	Uton	SX827988	2xtimber	8½
Sequers	SX632518	3x15	16	Uffculme	ST069126	4x16	8½
Slapton	SX828443	2x9 (333 o/a)	11				
Shinners, Dartington	SX 787621	10	12	Ventover, Teigngrace (VENTIFORD)	SX847747	12	11
Staverton	SX785636	5x21, 2x16	10				
Shipley	SX681929	16	9				
South Brent, New (AVONWICK)	SX715582	23, 8	10	Wrixele, Thrushelton (WRIXHILL)	SX465898	2x12	10
Spara, Ashton	SX843841	2x22, 7	9½	Ward, Walkham	SX541720	13, 10½	6
Steps, Dunsford	SX804883	3x16½	7½	Weycroft	SY307998	2x27	16
South, North Bovey (CLAPPER)	SX753828	14, 12	8	Withy, Broadclyst	SX975957	2x11½	10½
				Washburn, Halwill	SX797547	7½	9
Thorverton	SS935017	30, 40, 30	18	Wolford, South Bovey (WILFORD)	SX798797	21	8
Taw, North Tawton	SS656016	16, 4x13½	9				
Tamarstone	SS283055	2x13	10	Yeolm, Werrington	SX318873	2x18	11
Tinney, Lifton	SX393853	5x10	8½	Yeoford	SX783988	12, 11	8½
Tavistock, New (GUNNISLAKE)	SX433723	4x21, 19, 12	13	Yarty	SY281980	15, 21, 15	20
Tavistock, West	SX476737	3x16	10½	Yealm, Yealmpton	SX590520	12, 14, 12	11½
Two, Withycombe	SX608749	2x20	12	Yeo, Harberton	SX758594	13	15

APPENDIX B

LIST OF DEVON COUNTY BRIDGES 1831
(Original spelling)

BRIDGE	PARISH
Division of Black Torrington and Shebbear	
Woolley	Beaford
Beaford	Do
New	Huish
Hele	Hatherleigh
Hatherleigh	Do
Northlew	Northlew
Jacobstow	Jacobstow
Division of Braunton	
Linton	Linton
Bishopstawton	Bishopstawton
New	Tawstock
Winters	Do
Division of Crediton	
New	Newton St Cyres
Long	Shobrook
Shettern	Newton St Cyres
New Shettern	Do
Little Silver	Do
Lower Creedy	Do
Higher Creedy	Do
Codshead Pool	Crediton
Downes	Do
Fordton	Do
Uton	Do
Uford	Do
Gunstone	Do
Culverleigh	Do
Neapody	Do
Praydon	Sandford
Heath	Do
Preston	Do
Sandford	Do
Downridge	Do
Stockleigh	Stockleigh
Calves	Morchard Bishop
Division of Crockernwell	
Bridford	Dunsford
Sowton	Do
Steps	Do
Clifford	Do
Fingle	Drewsteignton
Dogger Marsh	Chagford
Rushford	Do
Chagford	Do
Paynes	Throwleigh
Ash	Do
Broad	North Bovey
North Bovey	Do
South	Do
Drakeford	Lustleigh
Wolford	North Bovey
Sticklepath	South Tawton
Metford	Tedburn St Mary
Division of Cullompton	
Thorverton	Thorverton
Elerhayes	Broadclyst & Silverton
Bickleigh	Bickleigh
Dart	Do
Exe	Morebath
Iron Mill	Bampton
Bampton	Do
Padbrooke	Cullompton
First	Do
Last	Do
Palmers	Do
Long	Do
Stoneyford	Do
Five Bridges	Do
Uffculm	Uffculm
Culmstock	Culmstock
Division of Ermington and Plympton	
Ermington	Ermington
Little Ermington	Do

161

Goulsford	Modbury	Cotleigh	Cotleigh
Sequers	Ermington	Uplyme	Uplyme
Yealm	Yealmpton	Axe	Axmouth
Torr	Do	Bridge adjoining Axe	Do
Puslinch	Do	Cullyford	Colyton
Laverton	Do	Sidford	Sidbury
Langham	Cornwoon	Sidmouth	Sidmouth
Parsonage	Do		
Harford	Harford	**Division of Lifton**	
Long	Plympton	Brightly	Okehampton
Plym	Do	Combow	Bridestowe
Shaugh	Shaugh	Stowford	Stowford
Bickleigh	Shaugh & Meavy	Wrixhill	Thrushelton
Lea Mill	Ermington & Plympton	Hayne	Stowford
Ivy	Ermington	Southern	Lifton
		Tinney	Do
Division of Great Torrington		Lifton	Do
Landcross	Landcross	Hele	Do
Iron	Do	New	Do
		Druxton	Werrington
Division of Holsworthy		Yeolm	Do
Blacktorrington	Blacktorrington	Poulston	Lifton
Diphard Mill	Bradford	Maristow	Maristow
Tamerstone	Pancrasweek	Coryton	Coryton
Bridgerule	Bridgerule	Gristone	Bradstone
Helscott	South Petherwin	Sydenham	Maristow
		Horse	South Sydenham
Division of Honiton		Lidford	Lidford
Finney	Feniton	Post	Do
Weston	Awliscombe	Belliver	Do
Town	Honiton		
Clapper	Do	**Division of Midland Roborough**	
Longford	Combrawleigh	Cadworthy	Sheepstor and Meavy
Up Ottery	Up Ottery	Sheepstor	Sheepstor
Awliscombe	Awliscombe	Lower Meavy	Do
Stoneyford	Combrawleigh	Higher Meavy	Do
Wilmington	Wilmington	Hoo Meavy	Do
Yarty	Kilmington	Hockworthy	Sampford Spiney
Bow	Axminster	Horra	Buckland Monachorum
Stoney	Do	Nosworthy	Walkhampton
Waycroft	Do	Ward	Do
Winsham	Thorncombe	Sticklepath	Buckland Monachorum
Long	Memsbury	Denham	Do

Division of Paignton
Fishacre	Littlehempston
Great Forder	Do
Little Forder	Do
Littlehempston	Do
Gatcombe	Do
Pords	Stoke Gabriel
Popes	Cockington & Paignton

Division of South Molton
Bow	Bow
Taw	North Tawton
Bundleigh	Bundleigh
Hansford	Chulmleigh
Head	Do
Brayley	East Buckland

Division of Stanborough and Coleridge
Dart	Buckfastleigh
Damarel	Do
Church	Do
Austins	Do
Emmets	Do
Leady	South Brent
Ball	Buckfastleigh
Shipley	South Brent
Dean Burn	Dean Prior
Brentharbertonford	South Brent
Brent	Do
New	Do
Glaze	Do
Skinners	Dartington
Yeo	Harberton
Washburn	Halwell
Gerah	Diptford
Lea	Harberton
Harbertonford	Do
Rosta	Do
Bickham	Diptford
Hatch	Loddiswell
Topsham	Do
Loddiswell	Do
Loddiswell New	Do

Aveton Gifford	Aveton Gifford & Loddiswell
Tacketts Wood	West Allington
Collapit	Do
Frogmore	Sherford
Ranscombe	Do
Bow	Blackauton
Bocombe	Sherford
Bow	Ashprington
Tuckenhay	Do
Old Mill	Ditsham
Blanks Mill	Malborough
Black Pool	Blackauton
Slapton	Slapton

Divisions of Tavistock
Abbey	Tavistock
West	Do
Higher Lumburn	Do
Shilly Mill	Do
Middle Lumburn	Do
Lower Lumburn	Do
Harford	Peter Tavy
Higher Hill	Mary Tavy
Lower Hill	Do
Trenow	Whitchurch
Kidnoll	Milton Abbott
Merryvale	Whitchurch

Division of Teignbridge
Doddiscombsleigh	Doddiscomsleigh
Crocombe	Hennock
Spara	Ashton
Bramble	Trusham
Heightly	Chudleigh
Kately	Do
Chudleigh	Do
Chudleigh Knighton	Do
Great Jews	Bovey Tracey
Little Jews	South Bovey
Drum	Do
South Bovey	Do
Benedicts	Bickington

Bow	Do	Ashclist	Do
New Bow	Do	Christow	Christow
Bow Ipplepen	Ipplepen	Cowley	Brampton Speke
Chipley	Bickington	Pynes	Upton Pyne
Aller	Kingskerswell	Stoke Cannon	Stoke Cannon
Dacca	Do		
Langford	Do	**Divisions of Woodbury**	
Keybury	Woolborough	Lawrence Clist	Clist St Lawrence
Lemon	Highweek	Clist Honiton	Clist Honiton
New	Kingsteignton & Ilsington	Bishopsclist	Sowton
		Grindle	Clist St Mary
Teign	Kingsteignton	Rockbeare	Rockbeare
Tuckers	Highweek	Fairmile	Talaton
Ventover	Teigngrace	Gosford	Ottery
Chirkham	East Ogwell	St Saviour's	Do
Morley	Torbrian	Dead	Do
Chuley	Ashburton	Cornercut	Do
Kings	Do	Cadhay	Do
Great	Do	Tailwater	Do
Gulwell	Do	Newton Poppleford	Aylesbeare
Cockmanford	Widdecombe in Moor	Otterton	Otterton
Dartmeet	Do		
Buckland	Do		
Exworthy	Do		
Pansworthy	Do		
Two Bridges	Do		
Dawlish	Dawlish		
Dawlish New	Do		

Total 276 Bridges

Compared with the 1809 list the following bridges were omitted from this 1831 list:
New, Tavistock   over the Tamar at Gunnislake
Staverton
One of the bridges adjoining Axe Bridge at Colyford
Holne
New, Holne
Newland, North Tawton

Division of Wonford

Minchin Lake	Heavitree
Topsham	Topsham
Ide	Ide
Kennford	Kenn
Heavitree	Heavitree
John Tolls	Kenton
Ashcombe	Ashcombe
Withy	Broadclyst
Broadclyst	Do
Burrough	Do

Recently constructed bridges, paid for by the county but not included were:-
Clyst Hydon   built in 1826, Woodbury division (?)
Newland Mill   built in 1827, North Tawton division
Wooleigh   built in 1828, Great Torrington (?)

APPENDIX C
SPECIFICATION OF WORKS IN REPAIRING BRIDGES 1831

DEVON COUNTY BRIDGES

Specifications of the several Works to be performed in Repairing and keeping in Repair, the Parapets or Guard Walls, and Railings, of the several Bridges of the County of Devon; together with the Roads over and adjoining the said Bridges, for Five Years commencing at Michaelmas Sessions, 1831.

1. The Parapet or Guard Walls, including the Coping thereof, and the Railings or other Fencing of the said Bridges, shall be immediately put in good and complete Repair, with the sort of Materials with which they have been usually repaired; the Masonry shall, where necessary, be Raked out and fresh Pointed with good Lime and Sand Mortar, and where any Coping is deficient, the same shall be made good. The Timber Rail Fencing, the Iron Sockets which contain the same, and any other parts requiring it, shall be painted in such manner as the Surveyor of the County may direct, in the course of the next Summer; and the said Painting shall be repeated once during the Term of the said Contract, at such time as the Surveyor shall direct; and the said Walls, Coping, or Railing, shall from time to time be kept in proper, perfect and substantial Repair.

2. The Contractor for these Repairs will be expected to keep a watchful eye over the state of the Masonry on all other parts of the Bridges which are not included in his Contract, and in case Repairs thereto shall from time to time become necessary, he is to give Information to the Magistrates appointed to superintend the several Bridges, and to the County Surveyor, of the precise nature of such required Repairs; and he is not to carry the same into effect until he has received an Order in Writing from the said Magistrates or Surveyor, directing him to do so.

3. Whenever any Bridge included in the several Lists, shall by direction of the County, be under a general or partial Repair, or Widening, Altering or Rebuilding, the sum allowed to the Contractor for the Repair of such Bridge or Bridges shall cease during the execution of such general Repair, Alteration or Rebuilding; and the Contractor will be required, on the completion of such Bridges, to maintain the Parapet, Walls, Guard Walls, or Railing of such latered or amended Bridges, in the state in which they are delivered to him, during the remainder of his Term, at the Rate or Price thereof stated in his Contract.

4. The Roads over the several Bridges, and connected therewith to the extent of a Radii of 300 Feet from each Abuttment of a Bridge, and further where it has been usually Repaired by the said County, shall be effectually Repaired; and where necessary shall be Broken up or Lifted, and Stoned with good, hard and durable Stone or Water Pebbles, broken so that no Stone exceed 6 ounces in weight, and properly divested of all Sandy or Earthy Particles, to the extend of the Hedges or other Fences on each side of the Road, leaving only, where it may be necessary, a sufficient Water Table; and such Stoning or Covering shall, in no case, be less than 10 inches thick in the centre, and 6 inches on the sides; and whenever it may be necessary to lay on new Material, the old Surface shall be properly Broken or Scarified, so that the new Material may properly unite therewith; and the whole must be so Raked in at proper times and seasons, that the Surface may become compact and smooth.

5. The Water Tables, Ditches, and several Drains, must be kept open and clean so that all Springs or other Water may be perfectly carried off from the Road; and the Hedges Fences must be kept in all cases properly Shorn, so that they do not project over the Road.

6. The Contractor is to find and provide all necessary Materials, at his own proper Cost and Charges, and pay all Damage done to Land or other Property occasioned thereby; but he will be allowed to exercise any power the Magistrates of the County may possess, for the procuration of such Materials. In case of any Alteration being made in any of the said Roads during the Term of the Contract, the same provisions are to apply thereto as specified in the case of Alterations in the several Bridges.

7. The Contractor will be expected to accompany the Surveyor on his inspection of Bridges, as often as he may deem necessary, on receiving notice to that effect; and on such Inspections the Surveyor will approve or disapprove of the state in which the Bridges and Roads may be; and when any Road or Bridge is found in an insufficient state, the Surveyor will refuse his Certificate for any Sum of Money which may be due on any Division of Bridges in which such Bridge is situate, until the whole of the said Division of Bridges shall be certified under the hands of Two Magistrates of the said Division; and should such Certificate not be delivered to the County Surveyor within One Month of the said Inspection, all Monies due on the said Division will be considered as forfeited.

8. The Tenders are to be made for the number of Bridges in each Lot, specifying, First – The Yearly Sum for the Repair of the parapets, Guard Walls and Railings – Second – The Sum per Annum for Repairing and keeping in Repair, the Roads over and adjoining each separate Bridge. The Amount will be paid once in each year, on the Certificate of the Magistrate of the Division and County Surveyor, viz, at or after the Michaelmas Sessions. Security to be given for the due performance of the Contract.

9. Should any doubt, dispute, or difference of opinion arise between the Contractor and the magistrates or Surveyor, touching the meaning or Intention of any part of this Specification, or the manner of Repairing the several Works, or Materials to be used, the same shall be left to the County Surveyor, whose decision shall be conclusive; and if at any time the Magistrates of any Division shall be of opinion that the Contractor is not proceeding in a proper manner in the execution of his Contract, or is not competent to do the same, they shall be at liberty to discharge him from his Contract, retaining such Sums as may be necessary to put the Works in a proper state of Repair.

JAMES GREEN
County Surveyor

Dated 1st September 1831
The Magistrates reserve to themselves the right to accept the Tender of such Persons as they may approve.

APPENDIX D

BRIDGE DESIGN AND CONSTRUCTION ASCRIBED TO JAMES GREEN

BRIDGE	YEAR	SPANS	LOCATION	COMMENTS

1808–1820

BRIDGE	YEAR	SPANS	LOCATION	COMMENTS
Yealm (at Lee Mill)	1808	1	SX599557	For Plymouth E TT
Fenny	1808	3	SY114985	
Drakeford	1809	1	SX789801	Widening only
Glazebrook	1809	1	SX690590	
New, Tawstock	1809	3	SS570282	
Yeoton (Uton)	1809	1	SX827988	
Cadhay	1809/10	3	SY093960	
New, Kingsteignton	1810	3	SX849764	Replaced 1845
Emmett's (Riverford or Hood)	1810/11	3	SX772637	Ashburton & Totnes Trust
Hele, Hatherleigh	1810/12	3	SS540063	
Uplyme	1811	1	SY328934	
Thorverton	1811/13	1	SS936017	84ft span since rebuilt
Langford	1811	1	SX871691	
Long, Membury	1812	3	ST255055	
Middle Lumburn	1813	2	SX454742	
Head	1813	3	SS667182	
Cowley	1813/14	3	SX907955	AM163 Joint city & county
Ash	1814	1	SX677905	
Pords	1814	1	SX857576	
Chudleigh	1814/15	1	SX857784	
Landcross Mill	1815	1	SS454235	
Colleton Mills (Hensford)	1815	3	SS664156	Widening only
Teign	1815	1	SX859734	Work also on approach arches
Steps (Dunsford)	1816	3	SX804883	
Bridford	1817	1	SX834872	Widening against 2 arches
Weston (Trafalgar)	1817	3	ST143001	
Hatch	1818	3	SX714472	Widening only
Culvery	1818	1	SX834985	AM270
Downes	1819	1	SX852995	Widening only
Rushford	1819	3?	SX705882	Widening only
Fairmile	1819	1	SX088971	Since rebuilt by DTp
Stoke Canon	1819/20	3	SX938975	Arch widening only
Dogmarsh	1820	1	SX713893	
Last	1820	1	ST027066	
Shaugh	1820	2	SX532636	Widening only
Brightly	1820	3	SX598974	Financed by Mr Savile, a magistrate.

1821–30

BRIDGE	YEAR	SPANS	LOCATION	COMMENTS
New, South Brent	1820/1	1	SX715582	Estimate £600
Clifford	1821	3	SX780897	AM 312, widening only
Clyst Honiton	1821	3	SX985935	Estimate £1200, widened by DTp
Sidmouth	1821	1	SY128878	For James Clarke
Metford	1821	1	SX817965	
Bassett's, Hatherleigh	1821	1	SS552035	
Dunchideock	1822	1	SX883877	Funded by James Pitman JP
Hele, Marhamchurch	1823	1	SS214037	For Bude Canal Co.
Canal drawbridge, Alphin brook	1823	1+1	SX940894	For Countess Wear bridge trust
Westwood, Stockleigh	1823	1	SS868030	Estimate £130
Great Huish Farm (underpass)	1824	1	SX82837	For Exeter TT
Beam Aqueduct (now pte road)	1824	5	SX473209	For Torridge Canal
Yeoford	1824	1	SX783989	Estimate £160
Eggesford	1824	1	SS683114	For Hon. Newton Fellowes

James Green – Canal Builder and County Surveyor

Name	Year		Grid Ref	Notes
Hynah	1824	1	SX836816	
Fulford Water	1824	1	SS996089	
Ermington	1824/6	1	SX640530	For Plymouth TT
Little Erme	1824/6	1	SX640530	do
Yealm	1824/6	1	SX590519	do
Gosford	1824/5	3	SY101970	Estimate £1200
Dawlish	1824/5	1	SX950767	Now called Stonelands
Chercombe	1825	1	SX833710	Estimate £250
Long, Cullompton	1825	3	ST025077	Private sponsors
Sandwell	1826	1	SX757598	For Totnes TT?
Long, Newton St Cyres	1826	1	SX881987	2 spans changed to culvert
Clyst Hydon	1826	1	ST036015	Estimate £300
R. Lemon, Bickington (Bow)	1826	1	SX794725	For Ashburton TT
Stoney, Honiton	1826		ST152012	Cost shared Honiton TT Demolished DCC 1968
South Bovey (Bovey Tracey)	1825	1	SX814782	Substantially rebuilt
Pillmouth (Landcross)	1826	1	SS695014	CI for Bideford TT, since rebuilt
Slapton	1826		SX828443	
Dart, Buckfastleigh	1827	3	SX744667	AM 145, widening only
New, Loddiswell	1827	1	SX719476	For TT
Otterton	1827	3	SY079832	Estimate £1700
Tinhay	1827	3	SX392851	Estimate £1200
Newland Mill	1827	1	SS659004	
Wooleigh	1828	1	SS522171	
Dipper Mill	1828	1	SS438065	£799
Kennford	1828	1	SX916864	
Sowden (Sowton Mill)	1828	1	SX823884	
Winsham, Thorncombe	1828	1	ST277060	For Alexander Hood (1824 in Somerset records)
Bampton	1829	1	SS959221	
Harbertonford	1829	1	SX784562	For Totnes TT
Bury	1830	1	SS738068	For Exeter TT with mill leat culvert alongside
Lapford	1830	1	SS727080	For Exeter TT
Little Dart	1830	1	SS668137	For Barnstaple TT
Newnham	1830	3	SS660174	For Barnstaple TT

1831–1841

Name	Year		Grid Ref	Notes
Countess Wear Swing Bridge	1831	1	SX940894	For Countess Wear Trust
Shuttern	1831	1	SX881979	
Belever Road	1832	3	SX658773	Constructed by Rev. James Holman, cost of materials only
Caton	1832	1	SX640545	Petition received
Wilmington	1832	1	SY218999	Cost shared Axminster TT
Winters	1832	1	SS579261	Cost shared Barnstaple TT
Polson	1833	3	SX355849	CI centre span since rebuilt. Cost shared Cornwall CC
Jews	1833	1	SX838764	Estimate £850. Since rebuilt
Barnstaple Bridge	1834	16	SS558329	4ft footways in cast iron
Tipton St John	1834		SY090917	Wooden, since rebuilt
Alphington	1834	1	SX917903	Since rebuilt 1960 floods
Loxbrook, Broadclyst	1833/34	1	SX995970	
Templeton	1835	1	SS877144	Estimate £300
Long, Plympton	1836	3	SX519567	£500 paid by Plymouth TT
Higher Creedy	1836	1	SS846012	
Crocombe	1836	3	SX848811	

Wrixhill	1836	1	SX465898	Demolished for DTp A30
Axe, Colyford	1837	1	SY259926	Estimate £2500 CI since rebuilt
Farrants	1837/8	1	SX836895	For Exeter TT
West, Tavistock	1837/8		SX476739	For Plymouth TT
First, Cullompton	1839	1	ST023067	Estimate £600, CI
Talewater	1839	1	SY092960	Estimate £160
Laverton (Lotherton)	1840	1	SX595538	Estimate £100
Cadover	1840/41	1	SX555646	Estimate £300
Northbrook	1840/1	1	SX939906	£200 towards Exeter TT
Newton Poppleford	1840/2	3	SY091898	Estimate £2500, Whitaker
Colyford	1841	1	SY254926	Estimate £500

APPENDIX E

JAMES GREEN'S CANAL AND DOCK PROPOSALS AND WORKS

ITEM	CANAL OR DOCK	SURVEY AND PLANS	CONSTRUCTION	REMARKS
1	Crediton		June 1811	Advertisement for canal cutters
2	Bude	1817 with T. Shearn	July 1819–1823	
3	Torrington	1810	1823–Feb. 1827	
4	Exeter	1818 to Oct. 1820	1820–1821	Dredge canal to 12 ft and repair exit lock
5	Bristol Ch. to English Ch.	1821 to 1823 June to Dec. 1824	= =	Tub-boat canal Ship canal with Telford & Capt. G. Nicholls.
6	Liskeard to Looe	Aug. to Oct. 1823	=	Completed by others using lock.
7	Newton Abbot	Dec. 1826	=	Short canal to town centre. (proposal only)
8	Exeter	1824	April 1825–Sept. 1827	Extension to Turf exit lock, 14ft draught.
9	Exeter		c. Sept. 1830	New basin at Exeter.
10	Exeter		c. Sept. 1832	Side lock to Topsham.
11	Grand Western	May 1829 to 1831	1831–1838	Ceased as Engineer 27 January 1836.
12	Cardiff Docks	1829–1830		Constructed by others.
13	Chard	1831–1834	1835–1842	Ceased as Engineer 1835, completed by others.
14	Burry Port		Opened 1832	Ceased as Engineer Feb. 1836 when walls failed.
15	Kidwelly and Llanelly	1832–1833	1835–1836	Ceased as Engineer. 30 January 1836. Inclined planes overspent.
16	London & B'ham	1833–1836		Spurred modifications to existing routes.
17	Newport Dock	1835	1836–1842	Entrance lock 202 x 62ft. Dock 795 x 240ft, with depth variable from 18 to 30ft.

APPENDIX F

MEMBERSHIP OF THE INSTITUTION OF CIVIL ENGINEERS 1838

President.

JAMES WALKER, F.R.S. L. & E.

Vice-Presidents.

WILLIAM CUBITT, F.R.S.
BRYAN DONKIN, F.R.S., F.R.A.S.

JOSHUA FIELD, F.R.S.
HENRY R. PALMER, F.R.S.

Council.

FRANCIS BRAMAH.
I. K. BRUNEL, F.R.S.
WILLIAM CARPMAEL.
LIEUTENANT DENISON, R.E., F.R.S.
JOSEPH LOCKE, F.R.S.

GEORGE LOWE, F.R.S.
JOHN MACNEILL, F.R.S., F.R.A.S.
W. A. PROVIS.
JAMES SIMPSON.
ROBERT STEPHENSON.

Secretary.

THOMAS WEBSTER, M.A.

Honorary Solicitor.

WILLIAM TOOKE, F.R.S.

Auditors.

E. J. DENT.   WILLIAM SIMPSON.

Treasurer.

W. A. HANKEY.

Collector.

G. C. GIBBON.

MEMBERS.

AHER, DAVID, Dublin.
ANDERSON, WILLIAM, Grand Junction Water Works.
ANDERSON, WILLIAM D., Newcastle.
ASHWELL, JAMES, Blaenavon.
ATHERTON, CHARLES.
BAIRD, FRANCIS, Upper Canada.
BAIRD, NICHOL, Upper Canada.
BALD, ROBERT, F.G.S., F.R.S.E., Edinburgh.
BARNES, JOHN, 14, Commercial Place, Commercial Road.
BAXTER, WILLIAM, Bangor.
BEAMISH, RICHARD, Daddershall Park, Aylesbury.
BIDDER, GEORGE P., 35, Great George Street.
BILLINGTON, WILLIAM, Wakefield.
BLACKADDER, WILLIAM, Glammis.
BLACKMORE, JOHN, Newcastle.
BLACKWELL, JOHN, Hungerford, Berks.
BLOM, Colonel FREDERICK, Stockholm.
BODMER, J. G., Bolton le Moors.
BRAITHWAITE, JOHN, 4, Adelaide Street, London Bridge.
BRAMAH, FRANCIS, Pimlico.
BREMNER, JAMES, Pulteney Town, Caithness.
BROOKS, W. A., Stockton upon Tees.
BROWN, NICHOLAS, Wakefield.
BRUNEL, MARK ISAMBARD, F.R.S., Thames Tunnel.
BRUNEL, ISAMBARD K., F.R.S., 18, Duke Street.
BRUNTON, WILLIAM.
BUCK, G. W., Manchester.
BUDDLE, JOHN, F.G.S., Newcastle.
BURY, EDWARD, Railway Office, Euston Square.
CASEBOURNE, THOMAS, Caledon.
CLARK, WILLIAM TIERNEY, F.R.S., West Middlesex Water Works.
CLEGRAM, WILLIAM, Gloucester.
CLELAND, JAMES, LL.D., Glasgow.
COLLINGE, CHARLES, Bridge Road, Lambeth.
COWARD, NOAH, Redruth, Cornwall.

MEMBERS.

CUBITT, WILLIAM, V.P., F.R.S., F.R.A.S., M.R.I.A., 6, Great George Street.
DAGLISH, ROBERT, Wigan.
DAVIDSON, JAMES, Stone, Staffordshire.
DONKIN, BRYAN, V.P., F.R.S., F.R.A.S., 6, Paragon, New Kent Road.
DONKIN, JOHN, F.G.S.
DRORY, G. W., Ghent.
DYSON, THOMAS, Downham.
EASTON, ALEXANDER, Market Drayton.
EDWARDS, GEORGE, Lowestofft.
FAIRBAIRNE, WILLIAM, Manchester.
FAREY, JOHN, 67, Guildford Street, Russell Square.
FIELD, JOSHUA, V.P., F.R.S., Cheltenham Place, Lambeth.
FORBES, Captain W. N.
FORDHAM, E. P., Dover.
FORSTER, THOMAS, Haswell, Durham.
FOWLS, SAMUEL, Northwich.
FOX, CHARLES, 28, Gloucester Place, Camden Town.
GIBB, ALEXANDER, Aberdeen.
GIBB, JOHN, Aberdeen.
GLYN, JOSEPH, F.R.S., Butterley, Derby.
GOODRICH, SIMON, Lisbon.
GORDON, ALEXANDER, 22, Fludyer Street.
GRAINGER, THOMAS, Edinburgh.
GRAVATT, WILLIAM, F.R.S., 7, Delahay Street.
GREEN, JAMES, Exeter.
HADEN, GEORGE, Trowbridge.
HAGUE, JOHN, 36, Cable Street, Wellclose Square.
HALPIN, GEORGE, Dublin.
HAMILTON, GEORGE E., Walton, near Stone.
HANDISIDE, WILLIAM, St. Petersburg.
HARDWICK, PHILIP, 60, Russell Square.
HARRISON, THOMAS E., Sunderland.
HAWKINS, JOHN ISAAC, Chase Cottage, Hampstead Road.
HAWKSHAW, JOHN, Manchester.
HICK, BENJAMIN, Bolton le Moors.
HOPKINS, ROGER, Plymouth.
HOPKINS, RICE, Plymouth.
HOPKINS, THOMAS, Plymouth.
HUNTER, WALTER, Bow, Middlesex.
IRVINE, Major.
JEBB, Lieutenant.

MEMBERS.

JONES, JAMES, 2, Great George Street.
JONES, Major H. D., Dublin.
JONES, R. W., Llangor.
KNIGHT, PATRICK, Armagh.
KOEHLER, HERMAN, Leipzig.
LAGENHIEM, G.
LANDMANN, Colonel, Greenwich Railway Office.
LEATHER, GEORGE, Leeds.
LESLIE, JAMES, Dundee.
LIPKINS, ANTOINE, The Hague.
LOCKE, JOSEPH, 1, Adam Street, Adelphi.
LOGAN, DAVID, Glasgow.
LOWE, GEORGE, F.R.S., F.R.A.S., F.G.S., 39, Finsbury Circus.
MACKENZIE, WILLIAM, Railway Office, Manchester.
MACNEILL, JOHN, F.R.S., F.R.A.S., M.R.I.A., 9, Whitehall Place.
MAUDSLAY, JOSEPH, Cheltenham Place, Lambeth.
MAY, GEORGE, Clackmacarry.
MEAD, JOHN CLEMENT.
MILLER, JOSEPH, East India Road, Poplar.
MILLER, JOHN, Edinburgh.
MILLS, JAMES, Battersea Fields.
MITCHELL, JOSEPH, Inverness.
MOSES, MOSES, Newport, Monmouthshire.
MURRAY, JOHN, Sunderland.
NEILSON, J. B., Glasgow.
NEWNHAN, THOMAS G., Newtown, Montgomeryshire.
OLDHAM, JAMES, Hull.
OLDHAM, JOHN, Bank of England.
PAGE, THOMAS, Thames Tunnel.
PALMER, GEORGE HOLWORTHY, New Cross, Surrey.
PALMER, HENRY ROBINSON, V.P., F.R.S., 2, Great George Street.
PARKES, JOSIAH, 21, Great George Street.
PEEL, GEORGE, Manchester.
PENN, JOHN, Greenwich.
PERKINS, JACOB, 21, Great Coram Street.
POTTER, JAMES.
PRICE, HENRY HABBERLEY, F.G.S., M.R.I.A., 4, Parliament Street.
PROVIS, WILLIAM ALEXANDER, 24, Abingdon Street.
RASTRICK, J. U., Birmingham.
RENDEL, JAMES MEADOWS, 34, Great George Street.
RHODES, THOMAS, Liverpool.

VOL. II. H H

MEMBERS.

RICHARDSON, JOSHUA, Newcastle.
ROBERTS, RICHARD, Manchester.
ROENTGEN, G. M., Rotterdam.
ROTHWELL, PETER, Bolton le Moors.
SAVAGE, JAMES, 31, Essex Street, Strand.
SEAWARD, JOHN, Canal Iron Works, Limehouse.
SIBLEY, ROBERT, 39, Great Ormond Street.
SIMPSON, JAMES, Chelsea Waterworks.
SIMPSON, WILLIAM, Pimlico.
SMITH, JAMES, Stirling.
SOPWITH, THOMAS, F.G.S., Newcastle.
STEEL, EDWARD.
STEPHENSON, ROBERT, $35\frac{1}{2}$, Great George Street.
STEVENSON, ALLAN, M.A., Edinburgh.
STEVENSON, ROBERT, F.G.S., F.R.S.E., Edinburgh.
STEWART, WILLIAM, Bordeaux.
STIRLING, JAMES, Dundee.
STOREY, THOMAS, St. Helen Aucland.
STUART, WILLIAM, Plymouth.
SUERMONDT, Y. D., Utrecht.
SWINBURNE, WILLIAM, Westminster Bridge.
TENNANT, CHARLES, 101, Upper Thames Street.
THOM, ROBERT, Isle of Bute.
THOMAS, JOHN, Highgate.
THOMPSON, JAMES, Lightmoor Iron Works, Shiffnal.
THORNTON, GEORGE, Wetton House, Daventry.
THOROLD, WILLIAM, Norwich.
TRUBSHAW, JAMES, Haywood, Staffordshire.
VIGNOLES, CHARLES, F.R.A.S., 4, Trafalgar Square.
WALKER, JAMES, F.R.S. L. & E., 23, Great George Street.
WELSH, HENRY, Newcastle.
WHISHAW, FRANCIS, 13, South Square, Gray's Inn.
WICKSTEED, THOMAS, Old Ford.
WILSON, ALEXANDER, St. Petersburgh.
WILSON, DANIEL, Paris.
WINSOR, F. A., 57, Lincoln's Inn Fields.
WOOD, NICHOLAS, Killingworth.
WOODHOUSE, THOMAS JACKSON, Loughborough.
YATES, WILLIAM, Birmingham.

GRADUATES.

Bazalgette, Joseph William, 9, Whitehall Place.
Beardmore, Nathaniel, 24, Dean Street, Borough.
Bray, William Bayley, Manchester.
Buckle, Christopher Richard, Tredegar Square.
Carr, Henry, 6, Great George Street.
Cooper, James, 23, Great George Street.
Cubitt, Joseph, 6, Great George Street.
Dixon, Charles, 9, Whitehall Place.
Ellis, George, 9, Whitehall Place.
Gordon, Lewis D. B., Thames Tunnel.
Gregory, Charles Hutton, Pimlico.
Handyside, William, 9, Whitehall Place.
Hays, W. B., Mill Street, Bermondsey.
Hemans, Willoughby, 9, Whitehall Place.
Heppel, John M., Mansion House Street.
Howe, William Weston, 4, Parliament Street.
Jeakes, William, 51, Great Russell Street.
Jermyn, George Alfred, 12, Golden Square.
Lynde, John James, 16, Great Queen Street.
Patrick, Thomas, America.
Renton, Henry, Bradford.
Rigaud, Richard, Pimlico.
Romaine, William, 7, Webb's County Terrace.
Routh, James, Railway Office, Euston Square.
Simms, F. W., 2, West Square, St. George's Road.
Stawell, Jonas, Doneraile.
Thomson, J. G.
Tompkins, W. G., 6, Paragon.
Townshend, Benjamin, Birmingham.
Tucker, John Scott, 21, Mecklenburgh Square.
Turnbull, George, Cardiff.
Turner, Thomas, Stamford Street.
Waterston, John James, Admiralty.

ASSOCIATES.

Abernethy, George.
Aitchison, George, 7, Muscovy Court, Tower Hill.
Albano, Benedetto, 12, Piccadilly.
Armstrong, John, Bristol.
Ayrton, Lieut. F., R.A., Bombay.
Baker, George, 2, Montague Place.
Baker, Hughbert, 2, Parliament Street.
Ballard, Stephen, Ledbury.
Bartholomew, Charles, Rotherham.
Barwise, John, St. Martin's Lane.
Bethell, John, 20, Mecklenburgh Square.
Bevan, John, 19, Buckingham Street.
Boorer, John, 3, Lower George Street, Sloane Square.
Borthwick, M. A., 23, Great George Street.
Bourns, Charles, Cardiff.
Braidwood, James, 68, Watling Street.
Braithwaite, Frederick, New Road.
Bramah, Francis, Jun., Pimlico.
Brandreth, Captain, R.E., Admiralty.
Brine, James Augustus.
Bull, William, Hostingley House, Wakefield.
Capper, C. H., Birmingham.
Carnegie, W. F. Lindsay.
Carpmael, William, 4, Old Square, Lincoln's Inn.
Cayley, Sir George, 48, Albemarle Street.
Combe, James, Dorking.
Comrie, Alexander, 23, Fludyer Street.
Cottam, George, Winsley Street.
Cottam, George Hallen, Winsley Street.
Crane, George, Yniscedwyn, Swansea.
Crawshay, William, 39, Upper Thames Street.
Cubitt, Lewis, Gray's Inn Road.
Cubitt, William, Gray's Inn Road.
Davidson, John Rankin, Stone, Staffordshire.
Davison, Robert, Brick Lane.
Dawson, Capt. R.E., 9, Somerset Place.
Denison, Lieut. R.E., Woolwich.

ASSOCIATES.

DENT, EDWARD JOHN, 85, Strand.
DEVILLE, JAMES, 367, Strand.
DONKIN, BRYAN, Jun., 6, Paragon.
DREWRY, CHARLES S., 77, Chancery Lane.
DUNDAS, JOHN F., Dumfries.
ENGLISH, HENRY, 12, Gough Square.
ERRINGTON, J. E.
EVANS, THOMAS, Dowlais.
FRANCIS, CHARLES L., South Lambeth.
FREEMAN, WILLIAM, Millbank Street.
FROME, Lieut. R.E., Chatham.
GREEN, JOSEPH, Exeter.
GUEST, Sir JOSIAH JOHN, Bart., M.P., F.R.S., F.G.S., F.H.S., 13, Grosvenor Square.
GUTCH, GEORGE, Bridge House, Harrow Road.
HALLEN, BENJAMIN, Winsley Street, Oxford Street.
HANDLEY, HENRY, M.P., 30, Pall Mall.
HARDIE, THOMAS GIRDWOOD, Blaenavon.
HARNESS, Lieut. R.E., Woolwich.
HEATHCOAT, JOHN, M.P., 8, Wood Street.
HEMMING, SAMUEL, Wolston, near Coventry.
HENDERSON, Lieut. Colonel, Wandsworth.
HENDERSON, PETER, Cardiff.
HENNET, GEORGE, 16, Duke Street.
HENRY, DAVID, Dublin.
HOLTZAPFFEL, CHARLES, 64, Charing Cross.
HORNE, JAMES, F.R.S., M.R.I.A., Clapham Common.
HOUGHTON, DUGDALE, Edgebaston, near Birmingham.
HOWARD, THOMAS, 7, Tokenhouse Yard.
HOWELL, JAMES, 1, Vincent Square.
HUNTER, JAMES, Bow, Middlesex.
INMAN, WILLIAM SOUTHCOTE, 57, Pall Mall.
JOHNSON, JOHN, Grosvenor Wharf, Millbank.
JONES, JOHN EDWARD, 22, Strand.
JOPLING, JOSEPH.
KENDALL, HENRY E., Jun., F.S.A., A.I.B.A., 23, Hunter Street.
KENNEDY, HENRY, New Street, Kennington.
KING, NICHOLAS, 2, Riches Court, Lime Street.
KNIGHT, SAMUEL J., Pimlico.
LAWRIE, ALEXANDER.
LEAHY, PATRICK, County Court, Cork.

ASSOCIATES.

LEATHER, JOHN THOMAS, Sheffield.
M'INTOSH, DAVID, 39, Bloomsbury Square.
MACQUISTON, PETER, Glasgow.
MANBY, CHARLES, 9, John Street, Adelphi.
MARSHALL, JAMES GARTH, 41, Upper Grosvenor Street.
MARTIN, HENRY, 49, Leadenhall Street.
MASSIE, JAMES J., Aberdeen.
MAY, CHARLES, 7, Cowper Street, City Road.
MILNER, JOHN, New Road.
MITCHELL, ALEXANDER, Belfast.
MOORSOM, Captain, Railway Office, Cheltenham.
MORELAND, RICHARD, 149, Old Street, St. Luke's.
MOSELEY, WILLIAM, 53, Great Ormond Street.
MURRAY, ANDREW, Millwall.
MUSHET, DAVID, Coleford.
NEWTON, WILLIAM, 66, Chancery Lane.
NICHOLS, NATHANIEL, Bethlem.
NICHOLSON, ROBERT, Newcastle.
NICHOLSON, WILLIAM W., Gray's Inn Road.
OLDHAM, JOHN, 9, Whitehall Place.
PAINE, JOHN DAVIES, 57, Lincoln's Inn Fields.
PARSONS, WILLIAM, Leicester.
PELLATT, APSLEY, Holland Street, Blackfriars.
PENN, JOHN, Jun., Greenwich.
PEPPERCORNE, FREDERICK S., 15, Buckingham Street.
PIM, JAMES, Jun., Dublin.
POOLE, MOSES, 4, Old Square, Lincoln's Inn.
REID, Colonel.
RENTON, A. H., Pimlico.
REYNOLDS, JOHN.
RICHARDS, JOSIAH, Rhymney Works, Monmouthshire.
RICKMAN, WILLIAM, Duke Street.
ROBE, Captain R.E. Ordnance Map Office, Tower.
ROWLES, HENRY, 15, Stratton Street.
SAUNDERS, WILLIAM WILSON, East Hill, Wandsworth.
SEAWARD, SAMUEL, Canal Iron Works, Limehouse.
SIMMS, WILLIAM, F.R.A.S., 136, Fleet Street.
SMITH, Capt. W. T., M.E., Madras.
SPEARMAN, J. M.
STANSFIELD, JOHN, Woolwich.
STEELE, THOMAS, Ennis.

ASSOCIATES.

STEPHENSON, R. M., 105, Upper Thames Street.
STUTELY, MARTIN JOHN, John Street, Adelphi.
SYLVESTER, JOHN, 85, Great Russell Street.
SYLVESTER, Professor, University College.
TAYLOR, THOMAS F., 7, Salisbury Street.
THOMPSON, ALFRED, Eccleston Street, Pimlico.
THOMPSON, JAMES, Glasgow.
TOWNSHEND, RICHARD, New Ormond Street.
TREACY, W. A., 9, Whitehall Place.
TREHERNE, EDMUND, 11, South Molton Street.
TREMENHERE, Lieut. Bengal Engineers, 33, Somerset Street.
VINT, HENRY, Lexden.
VULLIAMY, BENJAMIN LOUIS, 68, Pall Mall.
WALKER, JOHN, Maidstone.
WATERHOUSE, JOHN, Halifax.
WATKINS, FRANCIS, 5, Charing Cross.
WELLS, Colonel, R.E., 85, Pall Mall.
WHITE, GEORGE FREDERICK, Millbank Street.
WHITWELL, STEDMAN, Kentish Town.
WILLIAMS, CHARLES WYE.
WRIGHT, JOHN.

HONORARY MEMBERS.

AIKIN, ARTHUR, F.L.S., 19, John Street, Adelphi.
BARLOW, PETER, F.R.S., F.R.A.S., Royal Military Academy, Woolwich.
BEUTH, HERR, Privy Counsellor, &c., &c., Berlin.
BEAUFORT, Captain, F.R.S., F.G.S., F.R.A.S., Admiralty.
BREWSTER, Sir DAVID, LL.D., F.R.S. L. & E., F.R.A.S., Hon. M.R.I.A., Edinburgh.
CHILDREN, JOHN GEORGE, F.R.S., &c., British Museum.
COLBY, Colonel, LL.D., F.R.S. L. & E., M.R.I.A., F.G.S., F.R.A.S., M.R.A.S., Tower.
DUPIN, CHARLES, Mem. Fr. Inst., Paris.
FARADAY, Dr., F.R.S., &c., Royal Institution.
GILBERT, DAVIES, V.P.R.S., F.R.A.S., D.C.L., Hon. M.R.I.A., F.G.S., &c., &c., Gower Street.
GREGORY, OLINTHUS, LL.D., F.R.A.S., Woolwich.
HANKEY, WILLIAM, Fenchurch Street.
HERSCHEL, Sir JOHN W. F., F.R.S. L. & E., F.G.S., &c., &c.
LEGRANDE, Mons., Paris.
LOWTHER, Lord Viscount, F.R.S., F.S.A., 15, Carlton Terrace.
PARNELL, Sir HENRY, M.P., 19, Chester Street, Belgrave Square.
PASLEY, Colonel, R.E., C.B., F.R.S., F.R.A.S., F.G.S., Chatham.
PEARSON, Rev. W., D.D., F.R.S., V.P.R.A.S., Islington.
RICKMAN, JOHN, M.A., F.R.S., Duke Street.
ROGET, Dr., F.R.S., &c., 39, Bernard Street.
WALLACE, WILLIAM, M.A., F.R.S.E., F.R.A.S., Edinburgh.
WHEWELL, Rev. W., Professor, F.R.S., &c., Cambridge.
WILLIS, Professor, F.R.S., &c., Cambridge.

BIBLIOGRAPHY

Abbot, W., *The Turnpike Road System in England and Wales 1663–1840*, Cambridge University Press, 1972.

Binnie, G.M., *Early Dam Builders in Britain*, Thomas Telford, 1987.

Booker, F., *Industrial Archaeology of the Tamar Valley*, David & Charles, Newton Abbot, 1967.

Broadbridge, S.R., *The Birmingham Canal Navigation, Volume I, 1768–1846*, David & Charles, Newton Abbot, 1974.

Buckingham, W., *A Turnpike Key*, West Country Studies Library, Exeter, 1885.

Cherry, B. and Pevsner, N. (ed.), *The Buildings of England, Devon*, Penguin, 1989.

Clew, K.R., *The Exeter Canal*, Phillimore, Chichester, Sussex, 1984.

Cragg, R., (ed.), *Civil Engineering Heritage – Wales and West Central England*, Thomas Telford, 1997.

Hadfield, Charles, *The Canals of the West Midlands*, David & Charles, Newton Abbot, 1966.

Hadfield, Charles, *The Canals of the East Midlands*, David & Charles, Newton Abbot, 1966.

Hadfield, Charles, *The Canals of South West England*, David & Charles, Newton Abbot, 1967.

Hadfield, Charles, *The Canals of South Wales and the Border*, David & Charles, Newton Abbot, 1967.

Harris, H., *The Grand Western Canal*, Devon Books, Tiverton, 1996.

Harris H. and Ellis M., *The Bude Canal*, David & Charles, Newton Abbot, 1972.

Hawkins M.R., (Editor), *Devon Roads*, Devon Books, Tiverton, 1988.

Henderson C. and Jervoise E., *Old Devon Bridges*, Wheaton, Exeter, 1938.

Heyman J., *The Masonry Arch*, Ellis Horwood, Chichester, 1982.

Otter R.A., (Editor), *Civil Engineering Heritage – Southern England*, Thomas Telford, 1994.

Reader W.J., *Macadam: The McAdam Family and the Turnpike Roads 1798–1861*, Heinemann, 1980.

Ruddock E.C., *Arch Bridges and their Builders 1735–1835*, Cambridge University Press, Cambridge, 1979.

Skempton A.W., *British Civil Engineering 1640–1840: a bibliography of contemporary reports, plans and books*, Mansell, 1987.

Wallis A.J., *Dorset Bridges, a History and Guide*, The Abbey Press, Sherborne, 1974.

Webb S. and Webb B., *English Local Government, The Story of the King's Highway*, Longmans, 1913; reissued Frank Cass, 1963.

REFERENCES

CHAPTER 1

1 Green, James, Obituary, *Minutes of Proceedings of the Institution of Civil Engineers [Min. Proc. ICE]*, 1849–50, Vol IX, 98–100
2 Hadfield, C, *Canals of the East Midlands*, David & Charles [D&C],1966, 55, 62
3 Hadfield C. & Skempton A.W., *William Jessop, Engineer*, D&C, 1979, 52, 56
4 Broadbridge S.R., *The Birmingham Canal Navigations*, D&C, Vol I, 1768–1846, 30
5 Rennie, John, *Reports*, Vol 2, 22.6.1797 to 26.2.1801, 113–117, Institue of Civil Engineers Library [ICEL])
6 Rennie, John, *Reports*, Vol 3, 19.11.1800 to 4.12.1805, 438–444, ICEL
7 Tozer, C., *Account for services to the Lord Lieutenant of Devon*, 5 Sept. 1818, 1392, Add 2M, item E5, Devon Record Office, Exeter [DRO]
8 Rennie, John, *Reports*, Vol 4, 9.12.1805 to 29.10.1807, 112–118, ICEL
9 Rennie, John, *Report to the Lords of the Admiralty re Plymouth Breakwater*, 24.9.1806, West Country Studies Library, Exeter [WCSL]
10 Rennie, John, *Reports*, Vol 4, 9.12.1805 to 29.10.1807, 377–379, ICEL
11 Ibid, 21–23, 31–33
12 Lord Boringdon of Saltram, 'Account of embanking the Chelson Meadow in 1806', *Transactions of (Royal) Society of Arts*, 1808, Vol 26, 30–34
13 Worth, R.N., *Calendar of the Plymouth Municipal Records*, 1893, Fourth Folio. Misc. Papers 1800-1825, 245, WCSL
14 Teague D.C. & White P.R., *A Guide to the Airfields of SW England*, Devon, 1982, item 38
15 Buckingham W., *A Turnpike Key,* 1885, 20
16 George A.B., *Devon Floods and the Waterways of Bridges, Proc. ICE*, Part 2, 1982, 73, March, 125–134
17 Smith, Allen, *A History of the County Surveyor's Society*, County Surveyor's Society, Shrewsbury, 1985, 12,13
18 Tucker, D. Roy M.A., 'Quarter Sessions and the County Council Government in Devon in the Nineteenth Century', *Transactions of the Devonshire Association [TDA]*, Vol 84, 1952, 178–205

CHAPTER 2

1 Minutes, Devon Quarter Sessions, QS 1/23, 1/24, DRO
2 QS 87/1–2, DRO
3 45 Geo III c75
4 Stoyle, Ian, 'Messages from the Past', *Devon Historian*, 35, 1987, 25
5 Taylor P.T., 'An Account of some discoveries made in taking down the Old Bridge over the river Teign (etc.)', *Archaeologia*, 19, 1821, 308–13
6 QS 1/25, DRO, Michaelmas Sessions 1820
7 Whiting, F.E., *The Long Bridge of Bideford*, The Bridge Trust, Bideford, 1945, 8
8 Bideford Bridge Trust Papers, A1/C/1, North Devon Record Office, Barnstaple
9 Exeter Turnpike Trust [Exeter TT], ETT 3/2, DRO

REFERENCES

10 *Exeter Flying Post* [*EFP*], 13 June 1811, Advertisement for Canal Cutters
11 Diaries of Mr P.R. Webber
12 Deposited Plan [DP] 17, DRO
13 Reference 3704 Inclosure Papers, DRO
14 DP 46, DRO
15 Brushfield T.N., 'Enclosure of Land at Budleigh Salterton', *TDA*, 1890, Vol 22, 299
16 Green, James, Obituary, *Min. Proc. ICE*, 1849–50, Vol IX, 98–100
17 Rennie, John, Letter Book 7, 7 December 1811 to February 1814, ICEL
18 Ownership Documents, Imperial Hotel, Exeter
19 Buckingham W., *A Turnpike Key*, 1885, 26, WCSL
20 Cherry B. & Pevsner N. (ed.), *The Buildings of England, Devon* [Devon], Penguin 1989, 231
21 Jenkins, Alexander, *History of Exeter*, 1808
22 *EFP*, 6 June 1816, 3b
23 Alastair Penfold (ed.), *Thomas Telford, Engineer*, Thomas Telford, 1980, 7,8
24 Reed, Harbottle, 'Demolition of Ancient Buildings of Exeter During the Last Half Century', *TDA*, Vol 63, 1931, 273–82
25 Fursdon, David, *Fursdon, home of the Fursdon family*, Maslands, Tiverton, 1984
26 Devon, Penguin, 453
27 Rennie, John, Letter Book 7, ICEL
28 Acland, Anne, *A Devon Family*, Phillimore, 69
29 See Appendix F
30 Green, James, *Report on Alteration and Improvement of the Turnpike Road between Exeter and Plymouth, 30 June 1819*, Nettleton, Plymouth; Devon and Exeter Institution, Miscellaneous Tracts, Volume 52
31 DP, 1820, Plymouth and Exeter Road, 1,G IV, 21, House of Lords Record Office
32 DP 39, DRO
33 Harris H. & Ellis M., *The Bude Canal*, D&C, 1972
34 Hadfield C., *The Canals of South West England*, 1967, 142–164
35 Rendell, Joan, *The Story of the Bude Canal*, Stannary Press, Callington, 1987
36 Hadfield C., *The Canals of South West England*, 1967, 151
37 Buckingham W., *A Turnpike Key*, 1885, 21, WCSL

CHAPTER 3

1 *EFP*, 8 September 1825, 1b
2 *EFP*, 28 August 1824, 4c
3 *EFP*, 10 April 1828, 1d
4 QS 110/2, DRO
5 QS 105, DRO
6 de la Garde, P.C., and Green, James, 'Exeter Canal', *Min. Proc. ICE*, Vol IV, 1845, 90–113
7 Clew K.R., *The Exeter Canal*, Phillimore, Chichester, 1984
8 Harris, H. & Ellis, M., *The Bude Canal*, 1972
9 *EFP*, 18 January 1821, 4d
10 *EFP*, 18 July 1823
11 Hadfield, C., *The Canals of South West England*, 1967, 137
12 Goaman, Muriel, *Old Bideford and District*, 1968
13 Alexander, J.J. and Hooper W.J., *History of Great Torrington*, 1948
14 Hadfield, C., *The Canals of South West England*, 1967, 168
15 Ibid, 120
16 DP No. 76, DRO
17 Hadfield, C., *The Canals of South West England*, 1967, 39–41
18 *EFP*, 26 May 1825, 4B
19 Edwards R., 'James Green's Association with Cardiff Docks 1829', *Newsletter of the Panel for Historical Engineering Works, ICE*, May 1981
20 Farquaharson A., *History of Honiton*, 1868, WCSL
21 George A.B., 'Highway Engineering Achievements Exeter Turnpike Trust, 1820-1835', *The Devon Historian*, 45, 1992, 3–8

179

22 Report to the Trustees, 30 December 1820, DRO
23 *EFP*, 4 January 1821, 4c
24 DP No. 40, 30 September 1820, DRO
25 Buckingham W., *A Turnpike Key*, 1885, 24, WCSL
26 Exeter TT, Bills Book 1818–1832, DRO
27 Exeter TT, Order Book 2/11, DRO
28 Plan of Proposed Roads, Lyme Regis and Charmouth 1822, James Green, WCSL
29 Act of Parliament, 4 Geo IV c 10, 2 May 1823
30 Act of Parliament, 4 Geo IV c 109, 17 June 1823. Also see Perkins, Keith S, 'Lord Morley's Flying Bridge', *The Devon Historian*, 41, 1990, 15–20
31 DP No. 54, DRO
32 1148 M B4 21(iv)/11, DRO
33 Minute Books, Ref DD/BD/06, D/T/ilm 9, Somerset Record Office, Taunton [SRO]
34 DP No. 71, DRO
35 Documents, Chard Turnpike Trust, D/T/cha, SRO
36 *The Old Series O.S. Maps of England and Wales*, Vol II, Devon, Cornwall and West Somerset, Harry Margery, Lympne Castle, Kent, 1977
37 Buckingham W., *A Turnpike Key*, 1885, 25, 28, 30 WCSL
38 Pascoe J., *Existing and Projected Roads*, 1825, DP No. 68, DRO
39 Bailey C., *Projected Route Eggesford to Umberleigh*, DP No. 69, DRO
40 *EFP*, 4 July 1829, 1e
41 ETT 2/5, 10 February 1830, DRO
42 Barnstaple Turnpike Trust, Box 62, North Devon Record Office, Barnstaple
43 DP No. 94, 30 November 1830, DRO
44 Devon, Penguin, 1989, 288
45 Green, James, Obituary, *Min. Proc. ICE*, 1849–50 Vol IX, 98–100
46 Report Combemartin Harbour, North Devon Record Office, 50/11/62, 29 October 1827
47 Report Combemartin Harbour, *North Devon Journal*, 8 November 1827, 1
48 *EFP*, 22 April 1830, 3b, 3c
49 Woolmer's *Exeter & Plymouth Gazette*, Saturday 2 October 1830
50 Clew K.R., *The Exeter Canal*
51 Cox, Miss J., note to author
52 Plan in Dorset Record Office, Dorchester

CHAPTER 4

1 Rowe M.M. & Jackson A.M., *Exeter Freemen 1266–1967*, 1973, 26
2 *EFP*, 21 October 1830, 2c
3 *EFP*, 24 September 1840
4 *EFP*, 2 August 1832, 3a
5 QS 1/27, QS 1/28, QS 1/29, DRO
6 *EFP*, 6 January 1831, ed.
7 Rendel J.M., Obituary, *Proc. ICE*, Vol 16, 1856–57, 133–142
8 *EFP*, 21 October 1830, 3c
9 *EFP*, 14 April 1831, 3c
10 *EFP*, 6 January 1831, 3d
11 QS 1/27, DRO
12 QS 1/28, DRO
13 Bridge drawing, QS 88/17, DRO
14 Bridge drawing, QS 88/73, DRO
15 Cruse J.B., *The Long Bridge of Barnstaple and the Bridge Trust*, Aycliffe Press, Barnstaple, 1982
16 DP 104, DRO
17 Greenfield D.J., 'Silk Mills Road Bridge', *Somerset Industrial Archaeological Society Bulletin*
18 Hadfield, C., *The Canals of South West England*, 1967, 66
19 Hadfield, C., *The Canals of South Wales and the Border*, 1967, 38–40
20 Ibid, 104, 106
21 Wakelin, Peter, 'The Inclined Planes of the Kidwelly and Llanelly Canal', *South Wales I A Bulletin* No. 37, October 1984
22 Jones, Glynne R. & Morris, David, The Canals of the

REFERENCES

Gwendraeth Valley, *The Carmarthenshire Antiquary*, Part II, 1972, 29–48 and Part III, 1974, 83–96
23 Faulkner A.H., *The Grand Junction Canal*, D&C, Newton Abbot, 1972, 160
24 Hadfield, C., *Canals of the West Midlands*, 1966, 104
25 Map of Westmoor, Kingsbury Episcopi, DD/SAS/C212, SRO
26 Hadfield, C., *The Canals of South West England*, 1967, 66, 87
27 Green, James, Obituary, *Proc. ICE*, 1849–50, Volume IX, 98–100
28 Williams, Michael, *The Drainage of the Somerset Levels*, Cambridge University Press, 1970, 157
29 Ibid, 154. Enclosure Award, SRO 101, Act 1833 Award 1836
30 D/RA 19, SRO
31 ETT 2/5, DRO
32 ETT 2/6
33 ETT 2/6
34 *EFP*, 9 March 1837, 3c
35 *EFP*, 14 June 1838, 1b
36 Green, James, 'Description of Perpendicular Lifts', *Trans. ICE*, Volume II, 1838, 185–191.
37 *EFP*, 22 October 1840
38 Census on Film 19, WCSL

CHAPTER 5

1 Green, James, Obituary, *Min. Proc. ICE*, 1849–50, Vol IX, 98–100
2 Rowe M.M. & Jackson A.M., *Exeter Freemen 1266–1967*, 1973, 26
3 *EFP*, 7 January 1836
4 Byles, Aubrey, *The History of the Monmouthshire Railway and Canal Company*, Village Publishing, Cwmbran
5 Leonard T., *A Short History of the Newport Town Dock*
6 Miscellaneous Papers relating to the South Devon Railway Bill 1844, DRO
7 Garde, de la, P.C. and Green, James, 'Memoir of the Canal of Exeter', *Min. Proc. ICE*, Vol IV, 1845, 91–113
8 Buchanan R.A., *Nineteenth Century Engineers of the Port of Bristol*, Bristol Branch of the Historical Association, The University, Bristol, 1971
9 Copy from letter held by K.S. Perkins Esq.
10 Green, James, 'Account of the Recent Improvements in the Drainage and Sewerage of Bristol', *Min. Proc. ICE*, Vol VII, 1848, 77–84
11 Library of the Religious Society of Friends, Euston Road, London
12 *EFP*, 16 January 1851, 3a

ABBREVIATIONS

DP	Deposited Plans
EFP	*Exeter Flying Post*
ETT	Exeter Turnpike Trust
Min. Proc. ICE	*Minutes of Proceedings of the Institution of Civil Engineers*
TDA	*Transactions of the Devonshire Association*
Trans. ICE	*Transactions of the Institution of Civil Engineers*

ILLUSTRATION ACKNOWLEDGEMENTS

Institution of Civil Engineers. Drawings of canal lifts from Transactions Volume II 1838, (Pages 140-41) also the list of members of the institution in that volume. Drawing by J Green jnr of Exeter Canal as it would have been in 1566 from Minutes of Proceedings, Volume IV, 1845 (Page 74).

Clerk of the Records, House of Lords. Permission to reproduce parts of a drawing by James Green of proposals to improve the Exeter to Plymouth road, a document held in the custody of the House of Lords Record Office (Pages 49 and 90).

Devon County Record Office, Exeter. Drawings of Devon bridges made by Thomas Whitaker held in the custody of the Record Office. (Pages 13, 36, 114). Drawing by James Green of a 50 feet span cast iron bridge deck, held in custody of the Record Office (Page 64). Map of Exeter (Page 23).

Westcountry Studies Library, Exeter. Prints of Exeter House of Correction, also view of St. David's Church (Pages 46 and 71). Photograph of Cowley Bridge (Page 30, right) and Axe Bridge (Page 113).

Plymouth City Museum. Painting of Old Long Bridge, Plympton (Page 119).

Newport Library. General view of Newport Dock; also plan of the dock (Pages 146-47).

Cantrell, Dr. G. L. For permission to record details of a list of bridges of 1831 and of a maintenance contract of 1831 both originally published by the Clerk of the Peace of Devon

Greenfield, D.J. For article printed in the Somerset Industrial Archaeological Society Bulletin concerning the bridges of the Grand Western Canal (Pages 131-33).

Murless, B.J. For photograph of River Tone Acqueduct (Page 129).

Stoyle, D.I. For photograph of Thorverton Bridge, 1813, alongside the replacement bridge being constructed in 1908 (Page 27).

Dartmoor National Park Authority. Photograph of Steps Bridge (Page 34).

Devon and Exeter Institution. Prints of Bideford Bridge (Page 38) and Exeter Canal Basin (Page 76).

South West Water plc. Photographs of Exeter Canal Basin (Page 77) and Tamar Lake (Page 78).

Western Morning News. Photograph of Bellever Bridge (Page 117).

INDEX OF NAMES

Acland, Sir Thomas   17, 46, 54, 69, 85, 92-93
Aust, Surveyor of Post Office   31
Ayers, (of London)   72

Bailey, Charles   95, 97
Bingham, Reverend Richard   41
Boringdon, Lord (Lord Morley)   7, 11, 12, 40, 91, 151
Bowden, M.   35
Brown, Thomas   26, 35
Brunel, Isambard K.   5, 42, 144-145, 149-150
Buckingham, W.   52
Buller, Captain   101-102, 110
Bute, Lord   99

Carnegie, R.B.   14
Cartwright, Robert   39
Chapman, William   42
Chapple (County Surveyor, Cornwall)   120
Charles, Earl of Stanhope   50
Clark, W.I. (Magistrate)   110
Clarke, James   59, 125
Cleveland, Drake, Dyer & Webber   40
Collyns, Charles   88
Cornish & Son, Builders   23, 69
Criswell, Henry   14

Cubitt, William   69, 99, 104, 135, 145

de la Garde, Philip   144, 148-149, 152, 179
Dean of Westminster   151-152
Dike & Meyrick   145
Drake, J.E., Clerk of the Peace   127
Drake, Colonel   72
Drewe (Magistrate)   110
Duckworth, Sir John   110
Duke of Northumberland   50
Duke of Somerset   10, 26

Eales, Richard, Clerk of the Peace   18-19, 57, 70
Earl St Vincent   10
Easton, Alexander   53
Elton (Magistrate)   109

Fellowes, Hon. Newton   60, 85, 118, 124, 167
Fortescue, John Inglett   44, 45
Fowler, William   92, 135
Fulford, Baldwin (jnr)   72, 125
Fulford, Colonel   105, 108, 110, 112, 120, 137
   102
Fulton, Robert   50
Fursdon, George Sydenham   45

Gaunt, Thomas   134
Gidley, John (T.C. of Exeter)   137, 148
Goodwin, Joseph   39
Gordon (Magistrate)   110
Green, James (snr)   9
Green, Joseph   47, 104, 138, 143, 144, 145, 150
Green, Ruth   143
Greenslade, Robert   71

Hamlyn (Magistrate)   106, 109-110
Hanning, W.   93, 132, 136
Hare, William   123
Hippisley, Sir I.C.   70
Hood, Sir Alexander   61, 168
Hooper, Messrs   125, 179
Hull (Magistrate)   120
Hunkins, Messrs   97

Jessop, Joseph   42
Jessop, William   9
Jones, Thomas   101

Kekewich (Magistrate)   110
Kennaway, Mark   95, 120
Kennaway, Sir John   93, 110, 111
Kingdon, John   51, 137
Kitson, William   127

Langmead (Magistrate)   120
Lee, Messrs   22, 35
Lethbridge, Sir T.   84

McAdam, William   24, 56, 88, 136
Mylne, William Chadwick   42

Nicholls, Captain George   84, 170

Palmer, Henry Robinson   56, 100
Parker, Thomas   38
Pascoe, John   40, 95,
Perryman, John   35
Porter, Lt. Gen. George   42
Praed (Magistrate)   110
Price, Henry Hebberley   126
Provis, W.A.   130-131

Rendel, James   17, 47, 78, 85, 91, 106, 150, 180
Rennie, George   134, 144-145
Rennie, John   9-12, 16-17, 19, 30, 41-42, 84, 98-99, 127, 128
Robinson, Nicholas   100
Rogers (Counsel)   148
Rolle, Lord   39-42, 54, 60, 80-81, 84, 144
Rolt, L.T.C.   53
Rowe, James   39, 43

Savile, Mr   36, 167
Shearm, Thomas   50-51
Sibley, Robert   100
Sillifant (Magistrate)   105, 110, 124, 142
Smeaton, John   9, 50, 98
Sparkes, George   42, 100, 103
Sparkes, Joseph   103
Stokes, Charles   125

Index of names

Summers, William 94

Taylor, Colonel 72
Taylor, P.J. 20, 33
Telford, Thomas 9, 13, 18-19, 45, 53-54, 73, 76, 84, 94, 97, 99-100, 106, 134, 149
Tozer, Charles 10, 178
Tucker, Thomas 118

Underwood, G.A. 54, 70

Walker, James 104
Walker, Ralph 42
Whidbey, Joseph 42, 54, 98
Whitaker, Thomas 36, 38, 81, 114, 123, 144, 169
White, William 15, 178
Whitley, N. 40
Woolcombe, Harry 12

INDEX OF SITES

Aller Bridge, Kingskerswell   28, 62, 127
Alphington Bridge, Exeter   59, 88, 109, 119
Ash Bridge, Throwleigh   33
Avonwick Bridge   59
Axe Bridge, Colyford   113-116, 122

Bampton Bridge   67, 107
Barnstaple Bridge   4, 37, 105, 126, 151, 168
Beam aqueduct, Torrington Canal   60, 81, 167
Bellever Bridge   112, 116-117
Bickington Bridge   48, 62, 153
Bickleigh Bridge   15, 107
Bideford Bridge   17, 34, 37-38, 151
Bishop's Tawton Bridge   95
Blachford Park, Cornwood   54, 97
Bovey Bridge   62
Bow Bridge, Ipplepen   116
Braunton Marsh   17, 40, 42, 95, 152
Bray Bridge, South Molton   109, 119
The Bridewell, Exeter   22, 23, 69, 101, 124
The Bridewell, Newton Abbot   72
Bridgetown, Totnes   10
Brightley Bridge, Okehampton   36, 58, 157
Broadclyst Bridge   107, 109
Buckland House, Buckland Filleigh   17, 44, 52
Bude Canal   18, 50, 51, 52, 53, 60, 72, 78, 83, 84, 128, 131, 152, 167

Budleigh Salterton, land reclamation   17, 41
Bunker's Bridge, Chudleigh Knighton   35
Burry Port   105, 134, 135, 137, 170
Bury Bridges   95

Cadhay Bridge   26, 153, 155
Cadover Bridge   112, 123, 142, 157, 169
Calves Bridge   107, 155, 158, 160
Cardiff Dock   7, 72, 99, 170
Cast iron viaduct, North Street, Exeter   144
The Castle, Exeter   18, 22, 28, 57, 65, 69, 71, 72, 108, 123, 124
Caton Bridge   118, 168
Cattewater, Plymouth   10, 12
Chard turnpike road   28, 25, 54, 84, 94, 107
Chard Canal   84, 105, 132, 134, 136
Chelmer and Blackwater Navigation   10
Chelson Bay (or Meadow), Plymouth   11, 12, 42
Chercombe Bridge, East Ogwell   62, 168
Chudleigh Bridge   21, 30, 31, 156, 158
Clifford Bridge, River Teign   59, 156
Clyst Honiton Bridge   59, 155, 157
Clyst Hydon Bridge   62, 164, 168
Coach & Horses Bridge, Honiton Clyst   109
Colleton Mills (Hensford) Bridge   31, 85, 95, 167
Colyford Bridge   123, 142
Combe Martin harbour   54, 98

186

Index of Sites

Cotleigh Bridge   28, 94, 107, 162
Countess Wear Bridge   15, 25, 37, 38, 144
Countess Wear Canal Bridge   59, 88, 122, 167, 168
The County Lunatic Asylum   106, 125
Cowley Bridge, Exeter   18, 20, 21, 23, 29, 30, 31, 43, 56, 151, 155, 157, 164
Cranford Bridge, Broadclyst   109
Crediton Canal   17
Creech St Michael aqueduct   134
Crocombe Bridge, River Teign   112, 121, 156, 158, 163, 168
Cromford Canal   9

Dart Bridge, Buckfastleigh   48, 97
Dawlish Bridge   34, 52, 157, 158, 164, 168
Derridge Bridge, Morchard Bishop   110
Dipper Mill (Diphard) Bridge   66, 154, 162, 168
Doggermarsh (Dogmarsh) Bridge   36, 154, 167
Downes Bridge, Crediton   35, 107, 161, 167
Dunchideock Bridge   60, 167

Eggesford Bridge   60, 85, 95, 97, 118, 167
Ellerhayes Bridge   107, 153, 158
Elmfield (House), Exeter   17, 42, 44, 50, 52, 100, 102, 103, 136, 151
Emmett's Bridge   21, 16, 58, 107, 167
Ermington Bridge   54, 85, 155, 158
Exe Bridge, Exeter   25, 37
Exe Bridge, Morebath   25, 121, 153, 154
Exeter Canal   53, 54, 72, 73, 102, 103, 104, 106, 144, 148, 149, 152
Exeter to Plymouth road   18, 35, 48, 50, 52, 84, 95, 119
Exeter to Tedburn St Mary turnpike road   60

Fairmile Bridge   35, 164, 167
Farrant's Bridge, Dunsford   108, 109, 120, 121, 136, 169
Fenny Bridges, Honiton   12, 13, 19, 25, 26, 123, 151, 154
First Bridge, Cullompton   122
Fulford Water Bridge   62, 168
Fursdon House, Cadbury   17, 45

The Gaol, Exeter   22, 54, 69, 70, 71, 72, 100, 106, 108, 123, 134, 135, 137, 139, 170
Garah (Gara) Bridge   24, 152, 158
Glazebrook Bridge   26, 167
Gosford Bridge, Ottery St Mary   62, 155, 158, 164, 168
Gosport, tideway at Alverstoke   18, 41, 42
Grand Western Canal   19, 105, 127, 128, 130, 131, 132
Grantham Canal   9
Great Huish Bridge, Tedburn St Mary   60, 86, 167
Gunnislake Bridge (New Bridge, Tavistock)   107, 160, 164

Harbertonford Bridge   65, 67, 156, 157, 159, 163
Head Bridge, Chulmleigh   27, 31, 94
Hele Bridge, Hatherleigh   17, 26, 27, 30, 45, 51, 78, 80
Hemyock Bridge   34
Hensford Bridge (*see* Colleton Mills)
Higher Creedy Bridge   121, 154, 161, 168
Holne Bridge   107, 159, 164
Honiton and Ilminster turnpike road   84, 85, 93, 94
The House of Correction, Exeter (*see* Bridewell)

187

Hynah Bridge, Hennock   62, 168

Ide Bridge   164
Ilfracombe harbour   54, 98
Ilminster, Honiton to Wincanton survey   84-85, 98
Iron Bridge, Landcross (Pillmouth)   107, 162

Jews Bridge, River Bovey   118, 199, 163

Kennford Bridge   66, 164
Keyberry Bridge, Newton Abbot   28, 157, 159
Kidwelly and Llanelly Canal   105, 132, 133, 137, 170
Killerton House, water supply   17, 41, 46

Laira Bridge   17, 85, 89, 91, 92, 106
Laira to Ladydown turnpike road   89
Laira Bridge   17, 85, 89, 91, 92, 106
Landcross Mill Bridge   33, 167
Langford Bridge, Kingskerswell   28, 109, 164
Lapford Bridge   67, 168
Last Bridge, Cullompton   25, 35, 154
Laverton Bridge   123, 142, 155, 159, 162, 169
Lee Mill   25, 153, 154, 155, 159
Liskeard and Looe Canal   72, 81
Little Dart Bridge   67, 95, 97, 168
Loddiswell Bridge   62, 153, 156, 159, 160
London to Birmingham Canal survey   105, 135-137
Long Bridge, Membury   94, 107
Long Bridge, Newton St Cyres   52
Long Bridge, Plympton   112, 120, 153
Loxbrook Bridge, Broadclyst   119, 168

Lyme Regis turnpike road proposal   85, 89

Mill Leat Bridge, Otterton   116
Minchinlake Bridge, Exeter   107
Monkerton Bridge, Pinhoe   109

New Bridge, Holne   107, 164
New Bridge, Kingsteignton   27, 28, 66, 153
New Bridge, South Brent (Avonwick)   59, 92, 110
New Bridge, Tavistock (Gunnislake)   107
New Bridge, Tawstock   19, 25, 26, 31, 95, 152
New North Road, Exeter   42, 43, 106, 124-126, 136, 137
Newland Bridge, North Tawton   65, 106, 154, 156, 164
Newland Mill Bridge, North Tawton   164, 168
Newnham Bridge, River Taw   67, 95, 97, 167
Newport Dock   105, 142-145, 152, 170
Newton Abbot Canal   72, 83, 170
Newton Poppleford Bridge   12, 63, 65, 67, 110, 122, 123, 155
Northbrook Bridge, Topsham Road   108, 122, 123, 142, 169

Otterton Bridge   63

Padbrooke Bridge, Cullompton   107, 161
Palmer's Bridge, Cullompton   56, 161
Pillmouth Bridge, Landcross   63, 168
Polson Bridge, River Tamar   67, 112, 115, 116, 118, 120, 168
Pord's Bridge, Stoke Gabriel   33, 156, 159, 163, 167

Index of Sites

Pynes Bridge, Upton Pyne   24, 154, 155, 159, 164

River Frome and floating dock, Bristol   144
Rushford Bridge, Chagford   35

Sandford Bridge, Crediton   107, 109, 55, 161
Sandwell Bridge, Harberton   62, 92, 168
Sequers Bridge   121, 155, 160, 162
Shaugh Bridge (River Plym)   35, 155, 157, 160, 162, 167
Sheriff's Ward, Exeter   18, 22, 23, 28, 69, 71, 101, 123, 124
Shuttern Bridge, Newton St Cyres   107, 108, 116, 155, 160, 168
Sidmouth Bridge   59, 162, 167
Slapton Bridge   52, 156, 160, 163, 168
South Devon Railway Bill   42, 148, 152
Sowton Bridge, River Teign   65
St David's Church, Exeter   18, 44, 45, 47, 52, 152
St Ives Harbour   54, 96
Staverton Bridge   25, 107, 156
Steps Bridge, Dunsford   33, 156, 160, 161, 167
Stockleigh Bridge, Crediton   107, 161, 167
Stoke Canon Bridge   35, 153, 154, 159, 167
Stourton to Dudley Canal   9
Stourton to Stourbridge Canal   9
Stratton roads survey   54, 92, 93

Tailwater Bridge   120, 155, 160, 164
Tarr Bridge (River Yealm)   25, 154, 155, 160
Teign Bridge   21, 30, 31, 33
Templeton Bridge   112, 117, 120, 168
Thelbridge Ford Bridge, Sandford   109

Thorverton Bridge   21, 25, 26, 27, 28, 47, 107, 151, 154, 160, 161, 167
Thrushelton Bridge   121, 154, 160
Tinhay Bridge   63, 65, 168
Tipton St John Bridge   112, 120, 168
Topsham Bridge (River Avon)   35, 120, 156, 160, 162, 163
Topsham Bridge, Topsham   67, 107, 108, 120, 122, 164
Topsham to Exmouth turnpike road proposal   54, 85, 97
Torquay, water supply and other proposals   104, 127
Torrington Canal   17, 37, 38, 54, 60, 72, 80, 81, 102, 144
Trefusis Bridge, Grand Western Canal   129

Uton (Yeoton) Bridge   19, 26, 155, 160, 161, 167

West Charleton, Kingsbridge   12, 40
West Bridge, Tavistock   121
Western Ship Canal   72, 73, 83, 100
Westmoor, Somerset, land drainage   93, 104, 132, 134, 135, 136, 137
Weston Bridge, Honiton   22, 24, 35, 167
Westwood Bridge, Stockleigh, Crediton   60, 167
Wilmington Bridge   67, 116, 118, 162, 168
Winsham Bridge, Thorncombe   62, 162, 168
Winters Bridge, Chapelton   67, 107, 118, 161, 168
Withy Bridge, Broadclyst   35
Woolleigh Bridge, Great Torrington   164

Yarcombe Bridge   34, 94
Yealm Bridge   61, 91